POETIC
WONDERS
FOR BILINGUAL
READERS

Kong Shiu Loon

江紹倫

The Commercial Press

責任編輯：黃家麗

裝幀設計：麥梓淇

排　　版：肖　霞

印　　務：龍寶祺

Poetic Wonders for Bilingual Readers

著　譯：江紹倫

出　　版：商務印書館（香港）有限公司

　　　　　香港筲箕灣耀興道 3 號東匯廣場 8 樓

　　　　　http://www.commercialpress.com.hk

發　　行：香港聯合書刊物流有限公司

　　　　　香港新界大埔汀麗路 36 號中華商務印刷大廈 3 字樓

印　　刷：中華商務彩色印刷有限公司

　　　　　香港新界大埔汀麗路 36 號中華商務印刷大廈

版　　次：2024 年 1 月第 1 版第 1 次印刷

　　　　　©2024 商務印書館（香港）有限公司

　　　　　ISBN 978 962 07 4673 4

　　　　　Printed in Hong Kong

This book is dedicated to ALL MOTHERS.
The selfless devotion and gifts of mothers to their children is
the single factor for human greatness.

謹以此書敬讚天下母親

Preface

I present in this book poems I have written in English, my second language. Included are also selected poems by English poets with my rendition in Chinese.

Together, they show the beautiful characters of the two languages with diverse cultural richness. They also demonstrate the brevity and power of the Chinese language for poetic expression. The poems are unmarked by punctuations, a very special feature.

Poetry is creativity. They show the thoughts and imaginations of the poets with words depicting wishes and hopes. They are nourishing foods for our souls.

Quite a few of the poems in this book cherish the mother-child relationship which promotes the development of the meaning of life. In the words of Abraham Lincoln: "All that I am, and hope ever to be, I get it from my mother."

Bilinguals are persons who understand and use two languages. Their worldview is wider and richer. Bilingual poems open vistas and thoughts representing the fast and dramatic changes in today's global village. People who know Chinese and English may total to more than a quarter of the world population.

Some 1200 years ago the great Roman Emperor Charlemagne had the insight of the power of human language. He said: "To have another language is to process a second soul."

The great German scholar and poet Goethe said: "Those who know nothing of a foreign language know nothing of their own."

Wittgenstein the linguist said: "The limits of my language mean limits of my world."

Nelson Mandela who survived more than two decades of imprisonment on the strength of the poem *Invictus* declared: "If you talk to someone with a language he understands, that goes to his head: if you talk to him with his own language, that reaches his heart."

These insights urge me to publish this book. I hope it will provide comfort and enjoyment to readers in the mist of increasing social uncertainty and pandemic effects. Poetry appreciation will shift our attention to positive aspirations.

I began writing bilingual poems regularly in 1999 when I was 86 years old. It was a celebrative time. UNESCO declared March 21 to be World Poetry Day to commemorate Albert Einstein's 140th birthday. It declared: "the aim of supporting linguistic diversity through poetic expression ... and to promote the reading, writing, teaching and publishing of poetry through the world."

I have loved poems since I was a child living in rural China. It was a turbulent time of wants and struggles. But farmers sang mountain songs frequently, radiating joy and hope. Their emotional outpour represented the mark of human pride and majety.

By contrast, written poems are more than intuitive utterings. They communicate multiple meanings to be interpreted by oneself over time, or by others in different circumstances. They endure social change to enshrine messages of truth, kindness, and beauty.

We need to affirm that the highest aspirations of man are those closest to himself, and external aspirations must serve our inner development and satisfaction. To enjoy poems is the most touching way to do so, especially to read, write, or appraise poems in languages we know and love.

The 208 poems in this book are presented in four sections: (I) Personal and Social, (II) Nature We Belong and Behold, (III) Children and Parents, (IV) Poems by English Poets with Chinese Rendition.

I am by training and career a professor and social developer. My experience of eighty years has been grounded in many lands at times of dramatic change. My poems reflect these diverse experience and outlooks. My close friend and colleague at the University of Toronto, Conchita Tan-Willman, had read some of my poems. She is kind enough to say: "Your poems are like homilies buttressed and inspired by the scope and depth of your firsthand and vicarious experience with people, literature, music and culture. They are soothing, wonderfully formatted and engagingly articulated."

I earnestly hope my reader will find similar comfort and enjoyment reading this book.

Kong Shiu Loon
Professor Emeritus
University of Toronto

前言

本書所載的詩是我用英文寫的。英文是我的第二語文。

我自 1958 年到加拿大研讀心理和教育學以後，工作和生活都常用英文。因此，我寫詩用的是雙語人熟悉的日用英文。

詩為心聲，表述生活中的所見所感，所悟所望。我於近七十年間在世界各地教大學和生活，詩敘的自然是那廣濶時空中的世事人情與中英文字的特性和美妙。

當今時代是多變及危機四伏，不同國家民族必須密切溝通，應用中文和英文，互相了解和合作。

詩是意真情深的文體，聯合國教科文組織於 1999 年設定每年 3 月 21 日為「世界詩歌日」，紀念愛因斯坦 140 歲生辰，鼓勵全人類讀詩、寫詩和出版詩集，推動普羅大眾廣發心聲，實現和平。

我決定出版本書，適逢法國詩人 Annie Ernaux 於 2022 年榮獲諾貝爾文學獎。讚辭說她「以勇敢和高度敏銳筆觸，揭示個人記憶的根源、隔閡及時代集體約束」，實敘一個世紀的《悠悠歲月》。

我在多倫多大學教育學院教學時寫英文詩，經常請求專教英文的同事指正，一位數十年如一日的同事竟成為我的粉絲。她特別欣賞我讚頌母愛神聖，以及大自然的無限光輝孕育。她說：「你的詩反映你的廣濶經歷，以及你對文學音樂和文化的深知沉思，叫人讀了心靈安慰。」

我禱望本書讀者可有同樣的慰藉。

多倫多大學終身榮休教授
江紹倫

Contents

Section I
Personal and Social

Section II
Nature We Belong and Behold

Section III
Children and Parents

Section IV
Poems by English Poets with Chinese Rendition

William E. Henley

Section I
Personal and Social

Mother

Mother cherishes telling people and me
How amazing an experience she was carrying me in her tommy
For ten months she was nurturing a person to be
Imagining and curious what growth and development her
    child would be

My mother is the wisest person I have ever known
She shows me how to serve persevere and accomplish on my own
Her love for me is unfailing and unconditional and heavenly
She cared for me day and night with love and intimacy

My mother is solid and steady as our mother earth is strong
Tethering me to courage stability and keeping me safe and warm
She was patient seeing me try my first step in infancy
Readying herself with open arms when I stumbled into her lap easy

Mother often recount her thoughts and joy in my birth
How wonderful it was to hear my first loud cry on earth
It means the severing of dependence and the gain of a power to be
I know going my own ways is linked to keeping a bond eternally

Mother held my hand to school the first day
Telling me to make friends and to obey
She was proud with tears attending my university graduation

And solemnly advised me to be humble and diligent in service
   to our nation

My mother had no opportunity to attend school for literary study
Her abundant wisdom was acquired from productive activities
She lived a life of wars and famine and revolution realities
She survived them all allowing no conquest
All these helped her to live contentedly and compassionately
People praise her for her generosity and humility
When she passed away many people pray for her eternal peace
Being her son I love and respect her always in heart and soul
In summation she and I had lived in satisfaction whole and total

Great Mother

My mother was one hundred and two
When she was still trying to find things to do
She did call her grandchildren by wrong names occasionally
They felt her love and care so very warmly

Mother said she did not want any more responsibility
Accepting that her mind was uncoordinated recently
She enjoyed hearing any family member's achievement at
    work or study
And showed her appreciation and elation notably

Mother was born in the last century
Her birthplace was in a humble remote village
She did not have the opportunity for school education
But she lived by Confucian ethics and Buddhist compassion
She taught me to spare no effort and creativity working for
    productivity
And be courageous in helping people and animal in
    severe tragedy

Mother had an unusual life experiencing famines wars and victory
She endured much material losses and spiritual affliction quietly
She often said that failure and suffering should champion for new
    endeavours
Humans have the natural power to adapt and embark
    on new paths

Once when I was ten mother guided me to watch clouds fleeting
She told me white clouds could change into dark clouds storming
And cosmic changes are usual as changes in worldly life
I should learn from people who mount life changes to grounds high
Mother is blessed with this amiable quality to make lifelong friends
Despite continuous wars and separation they got united again and again
Their children were cared together to become brothers and sisters
Like nations of all peoples and cultures united harmoniously continuous
Mother and I had not often been together due to conflicting
    circumstances
But she always had something to teach me in our togetherness
Shortly before she passed away she show me how to exercise lying on
    her back
To stretch two arms high with fingers locked like a bamboo mat
Then turn the arms left in a wide circle flexing the muscles tight
Repeating the turns rightwards until one is tired
The last time I was with my mother I told her I practised her exercise
    every night
She kept telling me how great it was to have me and family in her
    prodigious life
Emphatically she expressed her satisfaction and gratitude
    before she retired
In the following morning I found her out of living breath
    wearing a smile
That was how my great mother gone
To yonder sphere where she would surely live happily on

Up Mount *Emei*

Up Mt. *Emei* peaks I climbed
Leaving all worldly concerns behind
Through forests of multiple kinds of pines
The air smelled flesh and fragrant so fine
In the early hours sutra chants sounded divine
Calling in the sun to rise in the pearly sky

I visited the *Emei* to pray in its many ancient temples
To ask Buddha to perch my late mother's soul in cosmic chateau
Before she died Mom wished her soul to rest in the sphere tranquil
The opening palm of Buddha can direct her the right way to go

For five days I prayed for Mom's resting peace in thirteen temples
Buddha's merciful eyes assured that no soul would lost in
    wandering behold
Every soul shall find a peaceful place to settle
Filial piety prayers do help kins in opposite worlds to connect continual

I stand this morning watching clouds swaying in the wind
And wonder on which cloud my Mom could be riding
In my mind's ears the sacred Aum syllables repeatedly chime
The soothing rhythm is a lullaby that had calmed my infant flights
I realize then Mother is at peace in Buddha's harbouring
Her voice is as comforting now as when she was living

Looking up the towering Buddha on Golden Peak I kneel
The halo of his ten heads looking at all directions radiate deep
  in my soul
I plead that I may meet my Mom in my dreams
And have Guanyin saved me whenever I am in trouble streams
When I stand up I feel the tiny touch of a large blue moth on my sleeve
I closed my eyes trying to feel Mom's sudden visit not a make-belief
The same sensation I experienced when I visited her tomb few years ago
The love of a deceased person does reveal in moths or dreams occasional

Language and Identity

Did you dream playing Bach's Chaconne with violin
And felt the deep sorrow Bach was experiencing
Did you imagine reading aloud Sagan's Bonjour Tristesse to an audience
The assembly appreciated your perfect French with big applause
Einstein says music is the language of the soul
Charlemagne says to have a second language is to possess a second soul

Language is power through which humans conceive and convey reality
Also language expresses beauty in words and poetry
You learn your mother tongue naturally at home and in school
You learn your second and more languages as powerful tools
To effectively communicate with people of other cultures and heritages
To immerse into the emotions and visions of people in our global village
So you expand your cognitive and imaginative capacity in scope and
    outreach
Your life becomes more meaningful and contributing to humanity
You might have once heard a pine tree sing
In orchestration with clouds and rivers beyond mountain ranges
You recall *Laozi* said the best voice in the universe is silence

A person can meditate and listen to his own soul in beautiful melodies
His relation with celestial forces and people is shaped by his ingenuity
Of the twenty seven hundred languages in use Chinese is so unique
With constructive forms and varying sounds 5,000 words in use are so
    intrigue

Chinese calligraphy is an infinite beauty in form and fancy creativity
Brush and ink paints infinite pictures dancing to equal Nature's
    choreography
Chinese words are best for poetry visually and in provoking imagery
They transcend grammar rules to allow poet and reader to
    share mutual feelings
Are your dreams in colour
Do you speak a foreign language in dreams to express feelings and ideas
Psycholinguists say colours and languages are pivotal tools for speech
    and wisdom
They express in words and pictures of intuition and freedom
In real life intuitive perception and expression are honest and
    noble notations
They carry our cultural heritage and personal character in vital
    importance   ‑

Confucius once stood on a river bank and realized the truth of
    human life
He mused intuitively how time passes incessantly days and nights
He then decided to engage in artistic activities instead of seeking a royal
    office
And resolved to teach and promote wise deeds for his nation and folks
Today we are grateful to be a part of the Chinese culture and civilization
To learn Chinese and other languages and value our wisdom and
    aspirations

Gratitude

Gratitude is what I feel and show
It's feeling humble and thankful in full
From God Buddha and all cosmic powers
Life is a gift it occurs not without a high endower

In a humble house a weather proofed roof
My door's threshold is crossed by friends and foe
Their smiles light up my way through life
I thank every blessing through my dimming eyes

Travelling makes me feel larger than life
Travails allow me to experience storms and ice
Up the mountain top sunrise and sunset invite me to kneel
Winds of love lift me up into cloud splendour whiffs

I learned to understand love and care by giving same to all lives
Compassion is shown by sharing people's sorrow plights
My heart has no songs I sing universal songs
By destiny we all harmoniously belong
I am eternally grateful having the opportunity and ability to grow
All earthly lives are blessed with the sanctity to grow as a whole

Dignity

Human dignity is same for all peoples
He who debase it is himself evil
To work creatively and care for others will generate your dignity
When people rich and poor are able to harbour dignity it is a just society

A person's dignity is his most valuable possession
He feels it best when other people regard him as their role model
Like the moon brightens the whole world in darkness
A person of dignity presents love and compassion for all people in
    distress

In our present world of ideological divides
Some politicians equate dignity with human rights
While human rights are inherent in all people undeniably
Dignity is earned by helping the poor and disabled to live in sufficiency
Remember to respect people of varying abilities regardless of
    success or failure
As long as they live purposely and work diligently their lives are valuable
Human beings are supreme because we know
When the strong is humble and the weak is confident they both
    live equally

Aristotle says a wise man has dignity without pride while fools feel only
    pride
Buddha taught it is better to be kind than to be right

Confucius says a good man is true to self and kind to people of all kinds
Perhaps there is no sanctity nor definitive entity for dignity
It stands real when individuals work honestly and contribute
    to community

Literature and Us

In my teens when I was attending junior high
My young niece and nephew were with me all the time
They nudge me to tell them stories sitting close to my sides
How the tortoise won the ten metre race leaving the rabbit behind
And the cunning fox tricked the tiger to be king of jungle in delight
Nancy was five and Matthew six
They enjoyed the stories in endless repeats

In graduate school I studied education and psychology
And learned how human development flourish from fancy stories
Literature creates stories of life in real and imaginary circumstances
To imagine and to create help children to expand their scope and beliefs
    to infinity

Literary narratives transcend space time cultures and races
They amend ordinary affairs of families and communities to
    wider territories
Readers can appreciate life's ups and downs in varying circumstances
And reflect to accept one's own trials with challenges in persistence
The function of literature is to discuss and denote issues of human life
Authors present the noble and evil aspects of individual life and society
    in passionate light

Whitehead says it is in literature that the outlook of humanity
    receives expression

Goethe says authors passionately believe in mankind's perfectibility
Dickens says a loving heart is the truest wisdom
Tolstoy says kindness and brutality and courage to care are common
    grounds of humanity
Hugo says there are moments when the body is fighting but the soul is
    on its knees
I read Hemingway's *The Old Man and the Sea* with enduring admiration
How a fisherman and a shark battle killings in the roaring ocean
And give in one with earned dignity and the other a meatless skeleton

On my eightieth birthday Nancy and Matthew kissed me big
    left and right
They brought two illustrated books beautifully inscribed
One says *Rabbit lost its race because of pride so blind*
The other says *The cunning fox can fool the tiger who forgot its own might*
Nancy told me she studied engineering hoping to contribute to better life
Matthew was now a Professor of Literature feeling confident in delight
All these confirm the mighty powers of literature in simplicity
Simple stories can move children and adults to use their minds and enjoy
    fond memories
Parents who share stories with children will help them grow up valuable
    and merry

Ah My Native Land

Winds through ten thousand pines sound musical
Green leaves on a thousand hills shine beautiful
Playful clouds create shadows from the sun
Waves of thriving rice on the plain a sight so fun
Birds roost in the midst of fine woods
Fishes glide in pristine shallows feeling good
Such heartfelt dreams end I linger
Awake I pine for my native land ever

The Mind

The human mind is wider than the ocean
It can reach yonder shores in an instant
The mind is deeper than sea
It fathoms the relation of I and me

The mind embraces humanity
It drives me to benefit the we
It enjoys beauty in immense splendour
Thru emerald pines amber sunset is seen in awe

The mind and heart nurture tender and vibrant kindness
Understanding poverty and misery with prudent zest
Comforting a grieving soul from heart breaking
It spares no effort in showing compassion and giving

The mind's power is immeasurable
It creates words images and thoughts in cosmic fold
Between introspection and outward wondering
The mind houses all comprehension
Yet for all the powers the mind has on tow
It fails to penetrate the fancy and mystery in its own fold
It could be the mind is in the head the heart in the chest
Challenged with the task of harmonizing East and West

What I Am

I live an ordinary life days and years
I listen to winds in silence with keen ears
I strive forward in varying paces toward my goal
I pay my dues treasuring happiness more than gold

In youth and old age I travel freely caring myself and others
I sit alone wondering how people love and hate one another
With my career and family I work to keep my community in harmony
In part and whole we weather the changing times successfully

Out of the hut of history oceans leap high and wide
At daybreaks with wonderous promises I rise with a clear mind
At times I greet myself at my own door to self-aware
And question my own mirror image asking who is there
I ask my friends and partners to take my hand and feel my thoughts
Along a wood path we together explore Nature's usuals and odds
Our desires and wishes ride on clouds to survey life's mysteries
We hear exploding bombs rather than gleaming thunders in decency

I remember asking my Mom long ago what is life
She told me life is a wonderful process of love and strife
To honour parents and tradition and to hold peace
To keep good company and contribute to humanity
There is no single cosmic meaning for all human lives
Each person creates his own life meaning like a poem fine

In later life I learned to be a livelong learner is to self-free
To develop and nurture growing relations with friends to be
To keep a leading role as I write the ventures of my story
That inner contentment and serenity are sources of sustaining happiness
That empathy and compassion for other lives foster thriving togetherness
To live with no regret nor debts of any kind and to leave a rich legacy
Life will then end with a peaceful mind in tranquillity
In unity with all cosmic elements my one-time journey is how life
    should be

Friendship

Like stern grass standing tall in shattering wind
Loyal friends feature open hearts happily wing
Human courage truly acts in interpersonal caring
Wisdom firmly dwells in mutual sharing
Loyal friends share robust deeds as well as imperfection
Their devotion overwhelms sad feelings and exhilaration

Friendship in Grassland

On this immense grassland of *Xinjiang* in mid-summer I stand
A carpet of emerald grass dotted by wild flowers extends to no end
In the unseen distance up the Sky Mountain white snow peaks glimmer
Even mighty eagles are shy to discover if animals for prey live there
Here I walk to no destination but to inhale the fresh air in motion
My horse grazes nearby enjoying the sweet and moist grass
    in full portions

Overwhelmed by tranquillity my mind is empty caring no worldly
    loss or gain
Grateful I am to have this discrete moment hearing silver bells ring
Where my eyes can see rows of tall trees form wide corridors for winds
    to blow
Far beyond a herd of cattle roams their thick wools echo white
    clouds in float
In a flash a rider with fancy attire swiftly passed my temporal sight
Beneath her elegant chapeau a gold and purple scarf tailed the wind
    swooping high

In the next instance the rider came back to greet me
In close proximity her big eyes silently asked me innocently
What could me an old man be doing in this remote place of wild
    grows only
I searched in awe for a reply to match her cares of simplicity
She understood my hesitation by throwing up her hands in extent

To show me the beauty of this vast place so pristine and immense
Then she introduced herself with a sweet smile her name is Alezo
And began to sing with her dombra strings this song from her soul

Hearing my song Sir you will be happy and forever remember me
My family does herding in this rich nourishing place totally free
We enjoy riding our horses and feel the winds from the mountain peaks
We set up our yurts for home in the most convenient and secure places
At night after a zestful dinner and feeling warm we watch the
    twinkling stars
My Mom would tell me which star I belong as I grow up to be
I dreamt of meeting this kind old man like Grandpa travelling
    in our territory
And sang my songs for him to remember me forever dear

Alezo's singing was so intriguing and touching I dozed off momentarily
    in peace
Now after thirty years I remember even my horse lay down on its sides
    at ease
The stars in the dark sky approved our heart-rending friendship and
    blessed me
I felt so safe with Mother Nature in her divine promise for fond
    memories so pretty

Big River and Me

My heart has always told me to visit the Yellow River
In early autumn when trees display their colours in splendour
Today I stand on the bank of the roaring flows
Remembering Confucius observation that time continues to go
Events of joys and sorrows were the makings of the Chinese people
The ups and downs of epochs and communities are narrated in words
    of simplicity
To show that family life is the foundation of culture despite wars and
    calamities
Kindness and evil deeds stand on trials of justice in wisdoms of divine
    morality

The Yellow River has been recognized as the mother river of China
Its muddy waters thru a vast territory had nurtured peoples in
    dynasties ever
Poet *Li Bai* sang the waters came pouring down from high sky
They flowed east to empty into the sea with no returning tides
Both Confucius and *Li Bai* observed that change is the nature and
    law of life
It challenges people to adapt and create meaningful lives
Adjustment is a moderate way of avoiding doing things in extremes
They feature interpersonal respect and cooperation in achieving
    goals foreseen

I used to teach at Toronto University the wisdom of Mid-way

and Constancy
A wisdom that had guided the Chinese and others for
    harmonious existence
In 1998 I revisited the university after a long retirement in Hong Kong
A professor greeted me with a warm hug telling me she was my
    loyal student
She has been teaching what I taught as fine lessons
To help her negative student to reflect and alter any harmful action
As a professor for twelve years she had been living a meaningful life
Reminding herself to be truthful and to care for others with gratitude to
    the divine

In a series of dreams I walk along the bank of the Big River in paces fast
    and slow
To realize how temporal and tiny I am in contrast to the river's
    mighty flow
Memories of my mother working in the field taking periodic rests appear
    in view
They tell me how common folks in China practise the Middle Way in
    times old and new
To care for self with good rests while working hard toward achieving
    mild goals
To be grateful for what Heaven gives and to share with those who
    received little

And stay far away from greedy and evil actions against humanity
  as a whole

While in sound sleep with dreams things happen
A gentle breeze held me up to ride on a colourful dragon to tour heaven
Looking down I saw a gigantic silvery snake travelling on nine twists
A leisurely stream meanders along a path with thirteen windings
The green pasture provides animals and birds a nurturing sanctuary for
  rest and foods
Above the distant snow peaks the sky features a pale green and crimson
  shine
In the horizon a setting sun seems too proud of its brilliance
  unwilling to sink
The cosmic forces orchestrate this grand symphony to accompany a
  day's yielding

If you wish to know how people on the loess plateau scope up muddy
  water for drinks
Just befriend a Chinese you will enjoy his respectful link

Happy Childhood

Happy are times of imaginative pretends
A stick can be a horse or an airplane
Five friends can form a toot toot train
Playing hopscotch requires no training

Happy is rising from bed alert and ready
Breakfast on table are foods of variety
Going to school is of no fearful concern
Time with friends and teachers is for intellectual discerns

Happy are timeless days of ceaseless joy
Singing and swinging are cheerfully deployed
Chums race to reach a distant tree
On arrival all laughing and dancing in gaiety
Happy at eve everyone share doing household duties
At ease after dinner Mom and Dad tell traditional stories
On moon and stars in the universe
On teachings of sages and beauties of poetry
Childhood is not a fixed entity
It is the beginning of a life journey

Happy Youthhood

Happiness in youth is marked with surging vitality
The growing young is not clumsy but full of energy
Vital powers are engines for productivity
They could also be endless doubts and destructive misery

Happiness is to feel for unlimited possibilities
Daring to learn traditional culture as much as novelties
To chat with friends on love and loneliness enhances social development
To day dream in solitude may lead to meaningful life experiment

Summer days are for fun and bravery
Sudden downpours and burning sun shines alternate swiftly
Times for me to climb up into the folds of the nearest tree
In minutes to swing a branch and jump into a river freely

Happiness is to wade along a stream in company of fishes
Water calms my soul fishes are delicious
School and university help to build knowledge and memories
Building a friendship of understanding and care enforces a loving life
    steady

Happiness is love of novelty and free trials working in concert
Having new clothes and doing on-line communication are normal sports
Times are for honouring parents and aging folks for continuity
To be filial is to love oneself as to care for community

Society should not expect youths to be leaders in future
Devoting and serving citizens must be nurtured

Happy Adulthood

Adulthood marks life's confidence and responsibility
Happy are adults fulfilling life goals successfully
Time to build up a thriving and happy family
Contributing to civilization and humanity

Happy is giving up fighting and negativity
And exalt your career and possibilities
Joy is inner feelings not related to outer circumstances
As the river of life flows joy flows swiftly

Happiness is universal harmony in measureless eternity
Personal failure and griefs harm our senses too deeply
Happy is to enable knowledge and skills to work for fitting achieves
Volunteer endeavours will enrich the lives of the needy

Happiness is sympathy and empathy shown in all circumstances
Achieving community prosperity and harmony in social victory
Happiness is pure rapture and joyous swell
Laughter is music for the soul in cosmic dwell
On evening respite watch the galaxies shine in darkness
With your family together be grateful to universal kindness

Happy Mid-age

Happiness in mid-age is graceful
It reminds that life runs in cycle
Times are reviewing life goals and accomplishments
Careers persisted in delight and rightful compliments

Mid-age life is staging dramas of autumnal gold
Bounty harvest in social deeds and natural growths
Children have grown in personal strength and social conscience
They cycle life forward for generational continuation

Happiness is making plans for active retirement
Beginning a second career and cherishing arts development
To heighten a spiritual platform for enlightenment
Happy are travels to appreciate world cultural establishments

Autumn life features receding roses their fragrance lurk at dusk
Falling leaves carry a thousand memories for new questions asked
Autumn is a season of love and contentment to behold
I dream while awake to see how withering leaves turning gold
When autumn meets tranquillity it is time to thank nature's blessing
Wild is the music of autumnal winds it tingle my soul ringing

Happy Senior

Happiness in seniority involves fears and eases
A little hard of hearing feels like you are enjoying perpetual peace
Free concerns of lack of money or dignity
Actively engaged in mentoring a child showing signs of ingenuity

Times are when you are looking to find your glasses
When in fact they are hanging on your neck fast
But seeing is real is no longer a dictum true
The mind's eyes see what is true more thorough

Happiness is doing volunteer work and charity
Caring the needy well marks the quality of a community
Some claim that aging leads to loss of human dignity
Watch a thousand-year-old tree standing you will see

Over the hill is not for you to say
'Cause you are on hill top busy with your gaze
When your knees are weak and in pain
Walk slow and often down memory lane

Senior years are considered to be golden years
Experience plus wisdom and humour invite no tears
Happy is learning to use the many operations of a computer
Information in words picture and music appear one after another
Happy is the senior who is shy of laments

How brilliant and comforting the sunset commands
As you approach the threshold of life's winter
You know people who believe caring for you an important task
Yours are hands that held your children high preventing any fall
The same hands will applaud their achievements large or small

Happy Old Age

Happiness in old age is marked by satisfaction and respite
No more struggle for excessive material or pride
Fears extent to physical fall and infirmity
A once vibrant body is now reduced to inability

No fascination relates to the invention of weapons for global annihilation
What more could add to the 1,550 nuclear missiles in powerful nations
Scientists with conscience do creations to benefit life natural
Happiness is to develop cures for virus and actions brutal

Time to lean back to enjoy tranquillity and nothingness
Happy to educate the young to love and be humble approaching
    progress
A wonderful world sees people of different cultures honour a shared
    aspiration
Cheerfully care and respect for lives on earth in harmony and
    continuation

Happiness is to play with children in open air
Energy and creativity apply to imagination so fair
Memories of bygone hardship and success presently show
Laughter and tear among loved ones calmly behold

Nothingness equates not to a life empty
In old age everything bursts with grandiose beauty

We experience riches and poverty journeying life with ambition and care
Happy to have satisfaction and meaning our adventures dared
Happy is to live with a partner each other understand
Overlooking a dripping jaw and hesitant move regarding them normal
Swallowing daily doses of pills with no choking is delightful
Happy it is to hold hands walking a journey final

Lying on bed with timeless leisure the old contemplates dying
A voice says in silence that death is rebirth for whom life has been
    winning
Hear larks sing praising rainbows over the oceans
Beauty is a joy forever and kindness in soul eternal
Life is a gift on narrow paths and broad roads
A good traveller leaves them wider for future generations to go

Laughing

Why is it important to laugh
Nancy more than once asks
Grandpa gives her big hugs as answers
In words and thoughts she giggles and again asks
To stay happy and healthy answers Grandpa
And be truthful and loving with all that matters
People who laugh often have good character
People who never laugh are of a different class

Why laughing helps people healthy Nancy asks differently
It expands lungs and blood flow Grandpa answers patiently
Could we sing songs about laughing alternatively
Let's do it rhythmically and loudly

I laugh on sunshine days or when rains fall
I laugh when I hear a mouse roar
I laugh working hard knowing leisure will follow
I laugh when your care wipes out my sorrow

I laugh when I found my lost bike
I laugh on mountain top after a long hike
I laugh when Mom bakes my favourite pie
I laugh dreaming my hair no longer white
I laugh when Grandpa cannot find me where I hide
I laugh when Nancy kisses me in surprise

I laugh seeing the moon tries to veil stars at night
I laugh when told the moon rises in a place not high
I laugh when I dreamed time is frozen not running
I laugh when my mistakes are judged as humour making

I laugh when my wrinkles appear more shallow
I laugh when my toys increase in multiple
I laugh watching world events like kaleidoscope views
I laugh at leaders cheat to feature daily news
I laugh when a cosmic voice says war and ills will pass away
I laugh knowing truth and love and beauty are here to stay

I laugh seeing Grandpa giggling unable to stop
I laugh knowing Nancy helps the needy in a volunteer co-op
We laugh when our world is resting in lasting peace
Let all peoples explode in joyful laughs passionately

Hope

Never let hopes go
Set an achievable goal
Pathways to success yours to find
Feathering hope to bravely fly

Life is both reality and dreams
Striving conquers both in gleams
Then happiness and satisfaction arrive
Riding a flying horse across the sky

Hope and goal are twins in our soul
Time and space help them grow
When meanings enrich our life in worldly settings
Aspiration and wishes make up our spiritual bedding
Then life is fulfilled
All blessings grateful

Fear

Neuropsychology regards fear as a good sensation
Even when it could be a perceptual distortion
When in fear our brain alerts us to make behavioural decisions
Either to attack or to stay calm to see clear the situation
Purposeful things done well are done without fear
In frightful circumstances we often give in to despair

Some fears are created by imagination
Like ghost in darkness or unknown situation
If you fear easily you should learn to strengthen your confidence
By knowing who you are and your aspirations
By widening your experience and appreciate beauty and novelty
To love and care openly and expressly

To speak freely and listen attentively
To feel deeply and act compassionately
To live adventurously and courageously
To lose serenely and respectfully
To risk wisely and daringly
To fear not time gone but time in eternity
To be what could possibly be

No one knows yet what fear is in reality or mystery
When in fear one runs faster or jump higher than usually
In lonely nights fear plays hide and seek with ye

On bright days fear may direct you to the wrong path in your journey
Ah how you wished someone is near as you fear the darkness
A fulfilling life must have companions in readiness

Life Worthy

As a child I asked my parents what life is
My Mom said my life is worthy as I make it to be
Growing up I realize life is much more than me
It has challenges physical mental and spiritual
To meet them effectively I have to strive and struggle
A noble task is to understand the person no one cares to know

Hardship and sorrow are inevitable in life
In surviving one should also have time and strength to be kind
Charity and compassion can shield one in sudden misfortune
Success in career and cares arise from love and knowledge confident
To live a life of decency one needs to share heart and soul
Everyone is a child of the universe no less than a star emerald

Life in cities today is full of noises and hustles
Craving for money and power exhaust every individual
We forget our cherished goal and spiritual ascension
Allowing ourselves shut in cages to slave for gold and position

To know who one is and how to be free
We can engage our mind to watch flowers and bees
And stay away from mundane sensory distractions
Breathing fresh air in open fields lost in beautiful observations
We are the earth the moon the ocean and everything
Living a life of dignity and benevolence is worth the wandering

Meditation

Let the attention in your senses and thoughts shift
To reside in your cool calm consciousness seat
Let your fancied love and casual improvisations
Off from the spring of your desire and passion
Put away your needs and greed
Aspirations and dreams for wealth and self-serving deeds

You will then ride while you are sitting
On the wings of silence
Across mountains of fear and worries
Seas of daily concerns and cares
To land on the quietude of paradise
A state of your mind
To realize a satisfaction in tranquillity
When your life is above survival
When you kindle what is meaningful
When you are your own master

Life and Death

My aunt ninety died recently after a lengthy convalescence
When friends and relatives offer their condolences
Her children reply that she was very old
As if it was just as well for her to go

A great grandson was born a week after she had died
There was jubilance to welcome him in a world of rights
In reply to people's call for celebration
The parents say the birth is just a new expectation
Both *Laozi* and *Zhuangzi* saw life and death as natural
Then life eternal is held in a person's own hands
Health is enhanced by being content of what life presents
And satisfaction deepens accepting the cycle of life and death

Only so would life begins its fruitful strives
And death dwells with a final peaceful respite

A Winter Dream

I pine for her violin
For her flutes and tambourine
For her orchestration of greens

I hear warm velvet breezes
Distant hill hisses
Bird songs shaking trees

I see tasty bamboos shoot
Buttercups and poppies vie for victory
The burst of flowers in variety

On endless sunny days
People lunch leisurely in shades
Memories of winter slip easily away
The beating of cold in abate

My dream takes a sudden turn
Ridding of beautiful thoughts unearned
Music of drums and cattle-runs rumbling
To remind me of this cold evening

A Dream

We bid each other adieu life unsettling even in a dream
Surviving calamities we meet again our hearts gleam
Let's treasure our love this wonderful moment
True affection appears in life how often

My Study

Layers of books and scrolls mark encounters memorable
This quiet small room beckons ideas and people global
On-line systems enable me to access the world in finger-tips
Reading playing music and chess are delightful acts indeed
Here I listen in quietude to heavenly tunes as clouds fly
Even space limits allow my mind to roam free and wild

My Humble Abode

Dwelling in a serene environment is my habitual liking
Carefree in quietude is enjoyment knowing not days passing
Washed by rain the spider net is clean and bright
Every spring swallows return to their homestead right

I often invite neighbours to share bamboo beauties
Calling my wife to brew cups of tea finest
Calendar dates and clock hours not in my interest sphere
In solitude I sit waiting for the moon to appear

New Year's Eve

New Year's Eve is romantic in words
It is also romantic in thoughts
Parting a person can hope to meet again
Parting a time the same hope has no end

The Luna New Year is celebrated by billions of people
Across space and cultures global
It is a mindful concern of people young and old
Recounting a year's events and plan for a new set of goals

In days old the whole night is spent cooking food for festive celebration
The culinary art of ten thousand years is golden tradition
Family members congregate in the busy and warm kitchen
The young learns how fine foods are prepared with love's sensation

At dawn firecrackers announce the new year beginning
Air and ground filled with red paper fragments lucky looking
From a distance drums and gongs roar loud
Led by playful lions and unicorns dancing proud
Everyone rush to the village square to join the gaiety
And exchange sincere wishes for peace and prosperity

Today we are in this social distancing world wide
Using the AI-phone for communication far and nigh
Any celebration is in silence for vigilance
Wishing our world rid of this deadly Covid-19 virus

Teacher in Action

What a teacher dedicates to do
She commits to bring to pass
Her teachings extends like drops of dew
To give moisture and life to every body and soul

A teacher is not an engineer
But a nurturing gardener
She controls not what and how children learn
But stays aside watching distinct personalities shone

My Teacher Friend

Those were years of social transition
When teaching was still an admired profession
Young men and women saw it as a lifelong career
Helping children grow and thrive was a challenging frontier

I watched Ms Ho handling a pupil sent to her for disciplining
He had wronged a classmate with swearing and spitting
Ms Ho asked and patiently listened to the pupil his story
He showed defiance and elegance telling her not to worry

It was a long counselling session lasting for half an hour
In the end the pupil recognized his wrongs with tears and a deep bow
Ms Ho was in charge of disciplining pupils other teachers
    could not handle
Soft-spoken and sincere she approached them to win their hearts in total

Teaching to impart knowledge is one easy job
To help children see themselves and to be free and good is another job
Children of the computer age require not information from books
They must learn to know self and others to develop perspective outlooks

I remember Ms Ho handling eight or nine difficult pupils every week
There were times when she and pupil came to terms together weeped
I asked once why she had accepted such nervy tasks so willingly
She said giving herself to teaching came only naturally

Anthony Ho at Eighty

If a teacher of eighty
Finds his students acting freely

If he recalls his years of toil
Matches the intensity of his pupils' joy

If a teacher is respected for his integrity
And finds himself in a picnic eating happily

If a teacher rejoices on his students' achievements
And receives a commendation by one in government

If a teacher hears from a forgotten student
To celebrate his affection long ago in a moment

If a teacher is loved for stern disciplining
By one who had committed a petit act of stealing

If a teacher retires not from running a blog
For the expression of feelings and ideas of his flock

If a teacher remembers not what he had taught
But sees his pupils' achievement in fields seemingly odd
Then it is the satisfaction of being a teacher
Who had never tried to be a preacher

Praise the Beautiful

Like a sparrow lost in the thick bush
Word recalled not however my mind pushed

Grandma makes best use of wasted materials
She boils a rusted nail in soup for mineral

If all poets sing the same tune
To save mother earth and humanity like tweens
And praise the stars and moon
The green pasture where cattle leisurely roam

Then we are not unlike the sparrow
Forgetting to let fly the imagining mind natural

Tranquillity Master

Tranquillity dwells in the garden of my soul
Each day I watch it fondly as usual
Green trees offer haven to bees and birds
They feed and nest while flowers burst

Tranquillity is the antidote for stress and worry
I am its master leading my life full and merry
After a busy frustrating time I meditate in serenity
My brain's alpha waves send away any lingering anxiety

Tranquillity thrives in acceptance and confidence
Life's interesting challenges could be in quiescence
Materials are givens and emotions emerge from interaction
Treat them caringly and harmoniously you will have satisfaction

Respond to difficult tasks and unfair accusations with a positive attitude
Apply creativity in your performance displaying gratitude
Rest and sleep with a peaceful mind
Tomorrow morning will be fine
The time will come when fame and riches mean little
We will return to mother earth in hope of an easy go

Life Path

How often do you try to find the right path of life
And it is in the distance you know not why
Deep in the forest when you trot on fallen leaves covering grounds
Sounds of footsteps ring musical notes up and down
The right path seems on air never could be found
But the echoes of woods are always memorable and divine
Finding the right path seems not as important as the find
In a sudden a ray of sun spears down through the dense foliage
The light points to paths dancing in steps with mountain springs in flow
You sit resting forgetting the trials of finding the right path to follow

In a dream you listened to Heidegger reciting his poem Path
    in the Woods
It was about bird songs and tree whispers in towering moods
You wonder how poets use words narrating such beautiful emotions
Lifting people up to cosmic sphere on sky ladders of imagination
You clap to show your appreciation in rhythms you never knew
Awake in a sudden you found the world is shining new

When I was a child my Mom told me following the right path of
    life is important
So I strive to do same with efforts in days and years wasting not
    an instance
Now in old age I recall all the success and failure gone with the wind

Along with the many winding paths trotted with heavy steps and
  on wings
My Mom's words are still ringing
So are tree whispers and bird songs chiming
My path of life is marked in a corner of Nature
I now sit resting contend in quietude and pleasure

Mementos

Of the trees on nearby hills
I like the pine the best still
They are just beautiful

My Grandpa treasures a prized memento
A pinecone as large as a pineapple
He received it from Stanford University's Provost
Her gift included two notes memorable

One tells the deed of a missionary long ago
He brought three young pine trees from west oriental
To grow in America making the west beautiful

The second quotes a Chinese proverb from long ago
In English the five-word verse is light gift conveys heavy affection

Grandpa's pinecone records the friendship of peoples
It is also the love sunk deep in his soul

Computer Games

The Computer
Its games hold
People young and old
Their attention and time controlled
By a minority of techno
Who knows not human freedom
Nor the import of life value
Only profit venues

Our Lives

I saw eternity last night
Through a jolly panel of high-rise
All serene lights only dimly bright
Up the hills time leisurely lies
I saw nothing driven in the spheres
A vast shadow as it appears
Would time await action in an on-line world
The impatience of people their lives hurled

Uneasy Dawn

A low sky
Black clouds lie
No winds
Nothing breathes

The park is empty
Up above the dark hills
A sharp ray opens the window
To shine in silence

The world is instantly white
A pair of pigeons fly
Their white wings define
Landscape and trees
As if to free
A new day to please

Good Morning

The sky east is white
Pink Clouds lightly glide
The sun rises to say Hi
Trees tall
Hills high
Mother earth everywhere shines bright

Green Power on Campus

Toronto in July One
I've come to join the fun
Drawn by waves of sound and noises
Amid screams jeers and demanding voices
For rights to differ and to expose

At dusk I walk by a campus
Learning place for universals
Now drowned in bums of electronic drums
There was once Green Peace to save earth and people
Now adds Green Space to encourage open intimate dazzle

Green is nature's beauty extravagance
Its diverse gleams a serene time marker
Trees and vegetables grow and wither in cycles
For eyes to see and please
For ears to hear and keep peace

Should one question how social progress proceed
Should one remain meek
Are there differences between humans and nature
Or life is but pumpkins and travels
Pumpkins grow in quiet open fields
Travels go with cars and boats
Campuses are never for screams and body exposes

In an evening otherwise simple
I look for greens in milliard shades
Time remains serene never in haste

My Life at Eighty

Today is my birthday I'm eighty
I think of my Mom giving me birth a deed mighty
It was at a time of war and sorrow
Everyday Mom said was a day borrowed
There's the usual jubilation of a new-born to the family
The hardship of bringing me up in scarcity was not a worry

The eighty years of my experience were quite extraordinary
Wars deaths dislocation and change required adjustments many
We lived through threats struggle and hope supported by sheer ingenuity
Under my parents' protective wings I had luck and freedom plenty
I learned more from practical challenges than schools which
    was intermittent
My aspiration and achievement carry the Chinese cultural
    imprints incessant

Venturing to the unknown had been experiences exhilarating
Like catching a fish in a running stream by hand a chance fleeting
Like learning to tame a calf for riding an enjoyment so exciting
Like mastering reading and writing a process mind-opening
Like grasping math concepts a struggle that must not be yielding
Like going abroad to study facing so much novelty that is testing
Like leaving the motherland watching its cultural destruction
    so disheartening
Like growing old and feeling unsure a life condition so unexpecting

Mother had gone with her belief in doing good

In the end she resolved that life is to promote a larger good

Our family had seen floods drought and the sudden seizure of all
accumulations

The only way to triumph against the atrocities is to rebuild life in
a new location

I dreamed of my Mom last night wishing me a happy birthday

She told me of her satisfaction having me despite difficulties in
stupendous ways

When I asked her of her secrets to serenity and longevity

She laughed aloud and told me to live honestly and enjoy beauty with
no worry

I remember my Mom taught me when I was eleven living in the village

Do work with your energy frequently it will give you continuing mileage

Seventy years of practice is enough to attest the truth of Mom's wisdom

As I find myself busy today with jobs large and small impacting
my bosom

My daily schedule begins with jogging and greeting people on the way

Then visiting the market to purchase fresh food best buy of the day

The times remaining is for reading and resting with music and
beauties around

The twilight hours are for writing and reviewing memories abound

I visited my cardio doctor yesterday to enquire about my heart condition

Results of techno-imaging enabled him to joke that I'd live on with
    no question
'Tis not often that doctors joke about a patient's health probabilities
But jokes often provoke thoughts and questions more than factual reality
So in part-two of my dream I asked Mom what it was like to live long
She told me to continue to work and play until no more steam
    comes along

It finally dawn on me that energy is different with human will
    nicknamed steam
Physical energy do things while the willful mind powers life seen
    and unseen
My Mom had cancer in the abdomen when she was ninety-six
She continued to help with family chores for years until she felt too sick
When it was time for her to die she asked how to find her way
    after death
Receiving a good answer she turned off steam and went without breath
I will continue to practise Mom's teaching for my remaining years
When my time comes I shall go in peace as she had in her 104th year

Experience in Wind
and Desert

On wide-spread wings winds howl
They bring memories telling not how
I once walked through sands to find my winds of life
Only to find myself on wings a joyful ride
Winds are scary in a storm
They gust in all directions blowing strong
Bleezes sweep gently through corn fields
I hold hands with Nature to laugh in gratitude

Winds whisper secrets of wisdom in special languages
In solitude I listen to understand the messages
Journeying thru a desert you may find your identity
Your personality shows as you recognize anonymity

Silence is enjoyed where silence be
Winds will blow abiding speeches for us all to hear
In desert you feel the vastness and stars in nights of beatitude
From distant mountain passes winds come carrying goodwill
They help us forget storms of yesterday to enjoy present songs
Home is nigh where harmony and honour live long

Dreary and Worry

Must Covid-19 and misery keep company
Let's enter the mine of human capability
To survey the ups and downs of history
And focus on what could really be

In the journey of diligent human strives
When vigil and cares exercise their might
To conquer virus in air and on mists
And send them to oblivion fittingly

We will then celebrate health and vitality in unity
In the grand halls of every country
Songs of victory
Sing loud and joyful melodies
For sustaining human bliss
To echo on psychic highland devoid of misery

Love Recalled

When old our veins are like frosty channels to a shallow stream
When feeble sparks don't fire us up even in dreams
We lie on a couch sleeping away hours they return no way
It's our earned solace far from our young fanciful days

Remember our intense love like firework illumining the sky
Our deep love amid colourful sea corals where fishes glide
When morning blights this land you and I clasp hands tight
Our love is steadfast despite how time resolutely fly

Love Speaks

When words of love speak volumes of affection and devotion
Love roots grow in fertile soils spouting sweet emotion
When your arms encircle me in every direction
Time sits still on clouds singing songs with orchestration

When our love resolves you and I drown ourselves in passion
We sing to fuel the fires of mutual care and admiration
Days and years march as we walk in ever slower paces
Until death extinguish our life flame we go far in no haste

Love Wishes

I dreamed of a woman wearing wings spreading wide
In flight she whirls and shines like golden wine
Her warm and tender touches are sheer delights
She whispers words of encouragement mine to keep for life

She invites clouds to carry me out of anxiety and worry
She is the clutch that keeps my body and soul steady
Locking eyes and smiles we dream not in diversity
Our family safe and stable we ride on a sleigh in space
Our promise of love forever holds however fast or slow life goes
Together we sail to the horizon where the brilliant sun sets low

Love's Powers

In the darkness of death love helps us to see
In the silence of sorrow love helps us to hear
In the warmth of memory love remains with me
In the solitude of being alone love fills my heart with waves of seas

Recalling precious moments and joyful hours love brings up a smile
Reliving successful deeds and humorous events love stands on my side
When in grief and despair do turn to love for comfort
When in heart ache with trembling hands call in love for support

I am the architect of my own fate
I dedicate myself to explore the extent of humanity's estate
In love of myself I shall build a life of health and happiness
For family and country I strive to enhance stability and cultural richness

For wisdom I shall work to empower our minds to conquer curiosity
For demising fear and brutality I promote understanding and spirituality
Will you dance with me in the wilderness
Shall we together sail thru the ocean's immensity
With the powers of love we keep our universe in eternal continuity
In love's pure soul we are children of cosmos' divine charity

Memory

Memory crests and ebbs in our mental web
Memory shows a mix of déjà vu and fleeting fads
Memory larks at dawn singing beautiful songs
Memory asks trees arousing curses on noisy gongs

I summon past events in resemblance of drama or reality
Days of hope and bliss were too brief for affinity
Grieving nights were not just loneliness for suffering
Forgiven sins and given blessings are in merciful offering
In my remembering protracted dreams wing on breezes and storms
They enrich my life with meaning stirring vital strives to prolong

Memory has a much blighter scene on spree
Friends and helping hands hurled me up to wheels on spin
Achievements seen and accomplishments stored inducing applause
Generations of healthy happy family members gather in the ancient hall

Life without recallable memories is a life of no celebration
Life in a wealth of memories is a true life of succession

Pride

Of all human emotions pride is most complex and controvertible
From feeling happy and satisfaction to being egoistic and remarkable
A secondary emotion like love envy and jealousy
Pride could be either negative or positive

An infant claims that he is good and lovable
He also seeks affirmation that his claim is tenable
Thru history sages and psychologists try to tell the intricacy
They forget children can use their emotions naturally

Pride and vanity are roaming weeds
They grow in heart's gardens wide and deep
Flourishing in all seasons they choke flowers in beauty
Diminishing selfless deeds while they sing majestic melody

Pride consumes a person's innate kindness
Blinding his eyes of sufferings and incapacity
Kindness are roses their fragrance soothes sufferers
Gifts grateful their songs green hills spectacular

When my teacher told the class that he was proud of me
Everyone applauded and shouted we too are pleased
I bowed and thanked my friends accepting their praise
Together we shared the pride and compliments for days

Back home at Xmas my parents welcome me showing their pride
I show them my works in literature and the received prize
Before sleeping that night I thank the gods for the grace bestowed
Deep in my heart I feel happy and humble with what I know

Pride is both a private feeling and an interpersonal sharing
One needs to be confident and efficient as well as daring
Shared pride is achieved by cooperation and respect
Communal success and creation arise from give and take

Let's surrender the temptation of being an idolater
The universe and human life manifest in many characters
Human knowledge is engaging much of it being untold
Poetry and sciences shine equally enriching our life and soul

Shakespeare's Vaccine

Celestial time space and events are transcendental
People of today often interact with those long ago
Shakespeare hears the leader of the most powerful nation is at ease
Unperturbed that 86 millions of his countrymen dead with Covid virus
He recalls Macbeth's cry False Face Must Hide What the False Heart
    Doth Know
Othello also rings at his ear that Reputation is an Idle and
    Most Imposition

Shakespeare values his own life and goes to get his vaccination
The nurse asks which arm he wants his injection
As You Like It the poet replies baring his left arm in action
When done the nurse asks if he feels painful
Much Ado About Nothing the poet replies delightful
The nurse smiled and asked how this worldwide disaster seems to him
Like *A Midsummer's Dream* the poet replied showing he is not in a dream

While waiting the poet opportune to ask the nurse in return
How long will be the quarantine end if required
The nurse chuckled with this reply *On The Twelfth Night*
Pursuing on the poet asks the nurse's judgement about this
    terrible matter
And what policy and actions the world leaders are doing to quell
    the disaster
The nurse flaunts and replies nothing but *A Comedy of Errors*

Perhaps the responsibility rests on us citizens the poet propositions
Laughing loud this time the nurse retorts To Be or Not To Be That's
    the Question
When it is time to leave Shakespeare thanked the nurse for her help
She smiled warmly saying *All's Well That Ends Well*

Reasons and Justice

Imagination is driven by emotion more than intellect
The richer the emotion the wilder it gets
Reasons and logic vanish driven by time
Warm feelings remain with the hearts in rhyme

With the passion and hot blood of youth I once hold
Of love and blind dedication crossed long ago
My reasons cried loud and rocked to and fro
To end up in riddles of life deeds punishable

What virtues books say I have no use at old age
Nor slogans shouted on printed pages
I know injustice when it is shown or in disguise
To fairness like a creek its high source I admire

Team Spirit

Human beings are gregarious and collaborative
Working in pairs or groups they sing and achieve
Feeling good and being productive humans develop thru evolution
They accumulate skills and wisdom conserving and in revolution

Evolving from Genus Homo to Homo Sapiens noble
Ancient humans took eon centuries to become body and mind able
Our brains become lager empowering learning and visions
Advancing from modern age to today's global digital communication

In real life teams pursue share goals and spiritual elevation
Teams succeed best when members help one another in work
    and emotion
Operation in teams is readily seen in a game of tug of war
When every member gives his utmost for the success of all

Today scientific teamwork could be performed independently
Hypothetical dreams are tested in laboratories isolated in time and space
Days and months go by individual researchers lonely sweat
Broad minded sharing can create synergy of findings thru digital nets
When a successful finding emerge all those involved rejoice as a set
Einstein said the value of achievement lies in the achieving
His simple words tell the truth of the team spirit in working

Ah Cooking

You are always fond to cook
Your Dad did good fishing with baited hooks
Fresh fishes taste divine when steam-cooked in haste
With chopped ginger and green onions on top given hot oil blaze
You were young then dwelling in the old country
When foods were prepared daily as family needed

Now you live in a cosmopolitan city
When cooking becomes an art to show off occasionally
Your kitchen stores spices and herbs many
Their individual and mixed aroma smells heavenly
Cloves nutmegs muster seed and pots of herbs in balcony
Chili fennel cinnamon turmeric paprika totaling thirty

You learn to use spices and herbs in cuisines of many countries
And develop your own culinary art to cook meat and poultry
Once you ate Chicken Cacciatore in Milan with full appreciation
Amazed when the chef said it was cooked for hours in gentle action
In your many recipes from your Chinese heritage
Chicken should be cooked briefly for tenderness and juicy taste

You begin by getting free-range chickens from your trusted butcher
With breast and thighs deboned and hung dry per your order
You marinate the meat with brandy soy sauce and crushed ginger
And allow the marinade to sit for an hour while you do dinner

Then you heat up the wok with a good spray of Canola oil
And brown up the meat on all sides using your skill on toil
Giving a light sprinkle of kosher salt brown sugar and corn flour
You stir in a cup of chicken stock and have the wok on cover
The dish will stay simmering for five minutes to complete

A second chicken dish is a unique savour
You rub the chicken with coriander salt all over
You stir-fly two cubs of black tea leaves in a heavy skillet with cover
Add a brick of natural brown sugar on top to help burn up smoke
Then spread out the chicken on the tea leaves to smoke cook
Keep the heat on for eight minutes to set in the fragrant flavour
Shake off the tea leaves from the meat to serve on an ornate plater
It is a heavenly treat when enjoyed with white rice and Sake in a warmer

Poem Writing

To write a poem is no task difficult
A poem comes to mind insights on tow
Other times verses arrive row after row
Or when you least expect a word pops up like gold

You need a theme like this poem's title
To jot down relative words and ideas is helpful
Do not just sit and imagine hither and thither
Better scribble freely with pen and paper

Ah Ha you scream when stanzas are written
You lean back for quiet moments on no reason
A brief pause and rest so earned is a sweet gratitude
Deep in your soul you enjoy a rewarding solitude
As you hold on your completed poem in good care
You read it soft and loud your music filling the air

Price for Progress

I know these hills well
Amid valleys and plains they are five and twelve
Villagers work the earth to grow vegetables and rice
Nature blesses the growths with rain and sunshine

I ran around wild and free for years before thirteen
Befriending every inch of the soil and greens
Catches of fishes and shrimps provided extra proteins
O what a life of stability and personal esteem

I left home to pursue progress at fourteen
Returning to review changes seventy years since
My hills had been flattened to build roads and factories
Seen are no happy children running in happy confident paces
People say economic growth is modern progress
Would happiness and spiritual growth be better conquests

Our Encounter

Incessant drizzles reign
'Tis summer again
Where hills stand proud
Verdant dazzles speak loud
A path is edged vivid
On earth and memory pit
Sometime in spring or winter
Ours was a sentimental encounter

Summer Rain

Once during a torrential summer downpour
I took shelter up the nearest tree
In vain my wish
I am wet like a fish

Through the leaf lattice I look out to the plain
Veiled by lacy mists evaporation from rain
Nature's exquisite beauty serene
Calm as a ripening olive in pale green

Out comes the sun fast as a shower
It shines like hundreds of mirrors
Reflections on paddy fields glitter
Tender rice shoots competing to mature

I took off my shirt to dry in the baking sun
Readying it for wear the rest of day
As I again crisscross the winding paths run
My bare feet feel the heating pain so gay

Summer Sounds

Crickets chirp in summer heat
Shakes the night's heart-beat
'Tis your snoring tunes
That interferes with my dreams

Sounds in the universe
Quiet moonbeams
Loud star twinkles
Silent melodies in orchestration

Middle Way Stability

A person stands in an erect position
He is in the centre of all four directions
He looks up and down moving his head
The middle of his body is in central stead

Confucius used these observations to formulate a wisdom
He named it the Middle Way for human relations and freedom
A caring person avoids excessive success or deficiency
He is happy in the middle contend to have propriety and dignity

Every person is born in a family and community of a cultural tradition
Thus endowed he grows up and strives to be his own person
He learns and acts with his abilities feelings and aspirations

Knowing himself he relates to people and Nature onward
    to his destination
Amid the many changes of time space and social forces he
    experiences life
Including challenges successes failures trials feelings darkness and light
All these shape up regrets and satisfaction grief and elation
In the end his life is full and meaningful or mundane in existence

For a person or society Mid-way is a vital practice to ensure stability
Extreme operations of any kind only lead to divisive wins and losses

Harmony is the foothold of humanity valued by peoples in
   world community
Mid-way and Stability constitutes the wisdom for human continuity
In diplomacy nations should forge cooperation and respect differences
Human-rights nests in the dynamic powers of Nature and just societies

Centuries later Mid-way encountered Aristotle's Golden
   Mean presentation
It is a whole with a centre plus two extreme points in opposing
   orientations
The Greek sage taught that virtues were essential marks of good persons
They include moderation courage and justice in knowledge and actions
Justice drives people to have friends and enjoy the benefits of
   collaboration
Aristotle contended that happy lives shine in differences and cooperation
Where participants respect each other's contributions with
   equal importance

Human wisdoms are narrated in languages of different cultures
   and peoples
Through congenial communication we can now share them in minds
   and actions
Favouring cooperation against wicked sanctions we can triumph in
   any difficulty

In trades finances illnesses pride envy selfishness lust fighting
    and dishonesty
To uphold humanity dignity nobility prosperity wisdom and
    eternal continuity
We celebrate and enforce the wisdoms of all seven sages in history
Inherent in them are habits and actions for excellence and tranquillity

Poets

The poet feels and thinks his days
To build beauties immortal not in haste
He concerns himself not with strident words
But with concepts and meanings people impress and rock

The poet bows before dews and sunrise
He sleeps dreaming how stars twinkle in great distance
He humbles hearing thunders shake up souls
And expresses gratitude when rains green the paddy fields

Murmuring Brook Remembered

I remember this winding green brook down from the hill
Fondly in my heart's eye and feelings still
I used to visit and wade on it almost daily
Its fishes fruits and other foods have kept me healthy

It runs from a spring on the rocks uphill
Through terrace fields in the valley cucumbers and beans filled
Tea plants line up the slopes watched by towering trees
Birds and bees compete for wild berries all free

Summer or winter its crystal green water feels just right
Where tadpoles shrimps fish and water insects happily thrive
Where random rocks divert the murmurs to singing crests
Scattered pebbles their faces smoothed in ripples breasts

The shallow flow passes places scenic and bright
Whence the wind sweeps through berry shrubs and pines
Detached leaves of various shapes and colours fall
To accompany the playful twirls whispering intimate calls

Happy are moments early at dawn
When I dash up the slopes to pick up fresh fruits earned
These bee-stunk persimmons are chosen for their juicy sweet taste

Thongs of ants hurry to reach them before me in a race

My footsteps alarm sleeping frogs they quickly leap to safety
The concerted sounds betray a waiting snake
Its luck gone breakfast is not yet made
My biggest joy in any day is to catch a fish or two
To supplement vegetable dishes and rice the daily food
Gifts of Nature are plentiful and free
Whoever strives to access them will his rewards be replete

Now I am living boxed in a space of cement and steal
With the rhythmic workings of Nature far from my view
I pine for the hills and rills their enriching spirits preside
Suffice to experience them in memory I am satisfied

Remembering My Love

Every music I listen I hear you and I together so fine
Now you appear only in my dreams saying no goodbye
How I wish once again I can hold the real you tight
In songs as they echo over hills and rills far and nigh

Scholars

Some scholars think what their neighbours think
They know what academic rules say is worth knowing
They edit and annotate what is plain
And despair when their writings are judged as not worth the ink

A true scholar builds knowledge on practical labour
His ideas charged by critics remain not wavered
While he defends independent thought with humility
He bows not to worldly rewards or injustice

A Good Day

I had a good time today
Feeling great
I visited a friend in distress
Helping to set his mind at rest
I cooked a meal for my family
Three dishes they ate happily

I've read this poem by Blake
His depiction of love and hate
Thus when I'm ready for bed
I feel peaceful having no regret
That this day is done
And 'morrow will begin a new one

A Happy Summer Day

The day is burning hot
To the beach we trot
With sunburn oil covering our skins
Mom and I lock hands to begin
Toward the coming tides we plunge in
To allow white froths flying
And our bodies floating
High above is the pale blue sky
Not a bird flying by
When our body and mind relaxed and free
It is time to run ashore quickly
To repeat such actions many times more
Until our legs say we need to restore
It is a happy summer holiday

A Moment of Love

You and I sat on a jumping horse in college gym
Our backs touched tight with no seam
We spoke with our hearts trembling
Exchanging silent words of love lasting

That moment occurred sixty some years ago
Time and space had changed manifold
I now still hear words of heart trembles
They reverberate in my soul songs ample

Always an Immigrant

We carry our racial features
Unusual cultural baggage
Pungent spices
Enliven foods of contrivance

We have to brave the stares
The showers of questions
Caring intended to help us fit
Into the grinds of a life that sits

Cares that are too heavy to accept
Our differences seen as inept
Must we subject ourselves to belong
To a society always ready to wrong

Anxiety No More

Tree branches dance in dazzles
Green grass happily nestle
Sparrows hop knowing seeds are on way
As dawn heralds in a new day
Keep your arms stretched high
Your looks in delight
Nothing could invite
Anxiety to get ahead from behind

Applause

Applause is a nonverbal and body communication signal
An appreciation and encouraging symbol
Such actions of praise and thanks involve body and soul
With intensity and persistence the applause affect the receiver total

I remember my teacher praised me for a good essay I wrote
She read out the piece in class and said it was admirable
My classmates gave me a standing ovation applauding long and loud
I bowed in humiliation feeling grateful and proud
In some circumstances applauding could mean protest
Pointing out wrongs and unfairness loud noises unrest
Thus applause can either be a positive or a negative communication
Its use represent human emotions in specific directions

Human life goes on with self and others in interaction
We need support and recognition from one another in action
Applause is not only a powerful show of appreciation in musical halls
In work place and at home recognition and praises work wonders for all

At Last Vaccines

Days rally in sunsets
Birds sing amid beaming trees disquiet
Children play in delight
Mom and Dad pray for dear life

Covid-19 viruses prey on body and soul
Despair and sorrows unfold
At last vaccines become available
A great tiding for sufferers consolation ample

Autumn Feels

I gaze on the perfect white moon in the dark sky so clear
My mind touches all friends around the globe far and near
Time is the Mid-autumn Fest when we all cheer
Wonders of the universe where humans hold dear

Like tennis balls silkworms await their transformation
Dangling cocoons ready to signal life in action
Suddenly they broke to fill the air with colourful moths
Wings dry they stretch wide to fly forth

Last night frost visited us to announce cold days near
Trees over hills and valleys their multicolours appear
Behold the scenic charm of my tangerine grove
Amid emerald leaves hang thousands of orbs in gold

Behaviour and Feeling

Babies cry when born
They smile when warm
Strangers meet their minds detect
They communicate not to suspect

Classical Guitar

By the sound
I found
The guitar you hold
Is Renaissance gold

Skillful fingers stroll
Plugging tunes new and old
Burning Spanish melodies chime
Flamingo dance tunes charm

Braid strings and cables
Embroider laces centuries old
From Baroque romance
To San Francisco

Music is mediator
Between reality and fantasy
Crossing boundaries of cultures global
To instill nobility in the soul

In meditation
I follow voices high and low
Wondering how
Music makes human beings whole

Clouds Perceived

My child's eyes see a cloud as a fluffy gigantic marshmallow
Passing thru mounts their open mouths waiting to swallow
Grand are nature's creations and revelations
Urging a child to exercise his teeming imagination

Clouds are vibrant beauties so colourful
They change in shape and tinting hues like miracles
Poets of diverse cultures concord clouds into time and space
To narrate thoughts and perceptions relative to unique taste

Clouds know well human affairs under their flows
They travel the universe watching as humans meditate and billow
Clouds chase winds to activate cosmic functions
They rain to green valleys and plains for foods and procreation
Caring clouds blunder and thunder to let humans know
All life activities must work harmoniously to persist eternal growth

Courage and Reason

Have no fear hold your head high
Let your mind acts free your imagination wide
Let out your words present sensible and vivid narrations
Speak your inspiring thoughts and unlimited aspirations

Strive to work in fields and marshes regardless of distance
Wade on mountain streams and rivers of empathy and reason
The celestial forces have endowed you with a powerful intellect
Learn and use your brain power over time leaving no regret

Life's ups and downs are happenings normal
Success and failure in acquiring riches and fame are usual
Be courageous in opening new paths and explore truths with vision
Be grateful to Mother Nature as you experience humanity's mission
When the time comes for sunset as for the end of life
Your mind and heart will be in satisfaction your soul rises high

Desk

Before me is my desk
Wood grains show the beauty of forest
Tree roots grow branches of books
Their ideas and wisdom toot
Deeds of love and hate
Within the memory gate
Embracing strives and hopes
Feelings of joy and sorrow
To make life interesting
People and trees everlasting

Dreamland

No one appeared in my dream
A desert seen
Sand dunes and purple boulders
Lie side by side in terrains bold

Rapid-eye movements keep sight
Of opaque hazy sky
No birds fly
Where plain is wide

Horizon in silhouette
Against a distant sunset
On a giant silver-gray paper
The universe registers no waver

Between the wilderness and me
Stood the observer me
The subject and the object
The dreamer and the dream

Dreams

I dream flying past the morning star
To reach hills and valleys where're
When time sits still in pondering space
Where territories stretch in unlimited ways

I am the master of free travels
In body and soul in rivers knowing no shallow
Where my mind roams carried by wisdom of old
Pitching no camp to muse over messages told

I meet *Qu Yuan* and Muse on my way to intellectual market
Not to drop dromedaries but to pilgrim on creative heads
Whence I cry out 'tis to hurry the caravan to gain speed
To leave the morning star behind to go on a free spree

And when a dream ends ordered by the morning sun
There is no worry the contents of it have gone and run
Wasted in rapid eye movements and wishes for nothing
Dreams are the nurtures of mind and soul far fetching

Occasional Dream

On Friday Mid-morning I fell into a day-dream
An extending grassland accompanied me as I loitered beside a stream
White clouds in blue sky followed us creating shadows fast and slow
Me and my environ aspired to reach a roaring ocean far beyond
    the meadow

I woke up in a sudden recalling the warmth and beauty of my
    subconscious sojourn
The endearing memory drove me to look forward to a weekend
    of no concern
I remember my childhood when Nature's beauties were in my picture
    book well drawn
Some pages had winds bringing songs sounding serene and loud from
    distant oceans
Such were the fairy tales my Mom often share with me in gentle
    loving narrations
In her affectionate presence I grew up happy and confident to believe
    life is constant

But Alas how rocket killings and Nature's wraths had barraged our times
    and communities
We now live in the darkness of Covid-19 plus Omicron and an
    obliterated humanity

Our existence has been thrown into the abyss of fear foolishness and
    subjection
An occasional dream might bring back hopes to save us from further
    stupidity

Fears Conquered

Fear not the grandeur of storms and lightening
Fear politicians who lie in public smiling
Fear has many facets challenging human understanding
Fears conquered enable us to be true and enterprising

In daily experience fear is aroused by sudden unexpected happenings
Of strange scenes or failures creating sensory and mental imbalance
Human beings have phobias of various kinds
When not subdued a person cannot normally function and shine

Life and Nature are full of unknowns in horror or in beauty
An open mind will help us to savour what is familiar or extraordinary
If you suffer from claustrophobia do avoid big crowds or a tight space
If you have acrophobia do not aspire to be a heroic flying ace
An effective way to conquer fear is to live and work in modesty
And sing out your concerns and anxiety tenderly or loudly
Achievement and failure have equal status in value finality
Risk and build on fears will enable us to live a life mighty

Life Is

Some say life is a predestined process
I say life is a creative process with fellow beings in lands and seas
Some say life is joyful in miracles and sad in crisis
I say success and failure alternates in cycles
Some say life is full of trials and being vulnerable is unavoidable
I say life is marching forward and challenges are manageable
Some say life is pleasures carefree enjoyment and gamble
I say life is responsibility family and growths plentiful
Some say life is boring doing things routinely day in day out till life ends
I say life is working meaningfully with accomplishments time well spent
Some say life is mechanical with work prescribed by institutions
    and traditions
I say life is voluntary open to courageous venture and valuable inventions
A poet said life is a mirage a dream and ecstasy in nullity
I say life is a bliss in beauty kindness and truth in sublimity
Beauty reveres in the splendour of towering mounts and roaring oceans
Kindness helps the wealthy and the poor equivalently potent
Truth is celestial human timeless and permanent
Indeed life is more because our hearts feel and our minds know
Life is language music wisdom hope and relations of harmony
So let us agree to make life its utmost and let each life be

Let me be restless and ascending in the presence of splendours

Let me treat power and gains as mundane and temporal
Let me be satisfied my soul gracefully nesting in time eternal
Let me travel in winding paths journeying to my destination floral
Let me truly possess things and honours in memory
Let me invite stars to twinkle as I rest with my family

Let me be curious and wonder what is in darkness and mystery
Let me do things not as I wish but as my conscience tells me
Let me regard each life as a flower with nectars for butterflies and bees
Let me sing songs of joy and satisfaction forgetting anguish
Let me hold glitters brightly in my hands to enable all lives to see
Let me draw my last curtain when I am ready regret free
Let mother earth and humanity continue with sunset and sunrise to be

Janitor

The janitor does his business
Like a detective in full alertness
His eyes walk all over
The room his world ever

He swabs and blots
And picks up dirt no matter what
He acts like a rehearsed musician
A nun with no ambition

He cleans and mops with dedication
Treating his job as a profession
Whence dusts settle and dirt gone
He sits to rest but not for long

Deep at Night

I look out of my window in the small hours of night
Before me stands a mount of trees thick and high
Adjourning are buildings of various heights
The sky mute looks down with opaque eyes

Nothing attempts to stir or to travel
Scattered lights flicker in faint yellow
They too seem to be in respite
Could this be just an illusion of mine

Three skyscrapers tower up especially tall
They stand still like unsharpened swords
In darkness dotted lights glimmer not at all
People are in slumber some even snore

Life is not restive in this city corner when time is still
In a few hours hustle and bustle the space filled
It was only yesterday when typhoon Kaitake swept through
Ah this calm and silky reticence are just temporary too

Silence asleep is neither a beauty nor peace
This passivity is a wonder in a world of unease
My mind tranquil cherishes only rest
I care little what might or might not be the best

Desert Message Eternal

The desert is tranquil
I hear sounds still
Gentle rhythmic steps of camels
Carrying bundles of silk and ceramics and people

Travelling thru a desert is an exuberant experience
Days are burning hot nights frigid cold
One feels small and fragile in the unending space
But not lonely despite the outward vacuity
Milliard stars twinkle like smiling eyes
Time stands and moves as the varying landscape dramatized
Artistic winds carved walls of sandstone into architectural wonders
Silent cliffs and narrow valleys invite healing meditation
These are showy activities in the cosmic stillness
Where man and nature ritualize their togetherness
Such calls of the Arabian Desert invite my family
For a get together there on the eve of 2022
From Canada, China, and Dhahran
To celebrate a long waited reunion
To escape the persistent pendemic Covid-19 global
And to show pure respect for the humanity of desert peoples

A most special experience imprints our soul at Jahol lkmah
Deep into the narrow cliffs a boulder franks the sky
On and around it exhibit etchings and inscriptions

The works of the Nabatacan ancients
Brave nomads who dwelled in this inhospitable environ
The markings show vivid fanciful writings and drawings
Etched some 2,600 years ago by hands of felicitation
They spoke to stars and would be travellers for communication
To connote affection and wishes in prayers total
Confident that their voices would be heard and felt eternal

I turned to my son Norman who arranged for our reunion
With much effort engaging flight schedules and accommodation
And conveyed a translated poem of one inscription shown
To evince the amazing humanity of desert peoples long ago
With sincere blessings conquering time and space
Confident that man are masters and songsters of all ages

The etched poem thus sang
I believe we are half way thru
Our camels are walking slow
On this spot we have a brief respite
I pray the heavenly powers on high
To see future travellers thru safe and bright
As we will to our destination no deny

Flames for All New Years

On days approaching the 2022 year end
Around Dubai Riyadh and AIUIA my family and I joyfully went
To appreciate the desert culture of ancient peoples so romantic
To feel the exotic space of the Arabian Peninsula so very big
Where many desert tribes steadfastly covered
Leaving us this wonderful heritage of brave human endeavour

From the helicopter we had a bird's eye view mighty
Sweeping the desert to appreciate its boundless grandiosity
Sandstone mounts dot the desolate land for variable beauty
Nature's crafty hands had artistic works done generously
Here and there ancient tombs present flashy fronts of grandeur
People here honour their deads with lasting extravagance

A lone wolf is seen on the vacant grounds looking for prey
The empty environ replies to its inhabitants offering negative says
We realize the desert welcomes no blaring crowds
It favours visitors who esteem serenity at all hours

Norman my son arranged an ideal place for retreat
In the mist of towering mounts around Dhahran it sits
The Banyan Tree Sanctuary with huge tents nesting lodgings for peace
Our large tent offers two bedrooms and a lounge for affinity
Outside at the back an open place shines a pool and chairs

Further towards the picturesque mounts are cozy couches around a
   metal dish
Four feet in diameter it is set on a steady rock bed
An open fire is blazing with Singapore cordwood carefully set

After dinner we loafed around this huge flame in the darkness
A gentle breeze sends the cracking sparks dancing like wavering leaves
To dialogue with twinkling stars with a language of peace

Our minds turn to the rich Arabian literature long ago
The power of storytelling in the *One Thousand and One Night* fables
The *Flying Carpet* tales so universally known
The brilliant *Golden Ore of Imrul Qays* in echoes global
The *Epistle of Forgiveness of Abu Al Maarri*
And the *Quran* of Muhammad soothing yearning souls

What if *Aladdin's Magic Lamp* now comes to our disposal
Norman calls with his cell phone to speak to his brother in Toronto
Raymond and his wife instantly appear sending greetings in loud bravo
Space and time transcended by modern technology so easy
And our family is now together around a fire in tranquillity
We feel so grateful to the real and literary creations including the Genie

Our fire is now burning warm and high
In the quietude I asked if I am hearing Auld Lang Syne

My daughter Lana presents a recorder playing the song
The fire is now in multicolour flames shooting high and flickering long
They echo the lyrics known by so many people all along
And we'll take a cup of kindness yet
For old time's sake (Auld Lang Syne)
Our family shouts Happy and Healthy 2023 eternal
And all future years flaming warm and bright also

Statue

On city square stands a statue
Representing a war hero
Its hand holding high
Pointing to the impartial sky

To what benefits for mankind
A little girl asks why
Its hand and head covered
In thick snow so cold

Aviation

Early in the morning
I look up the sky wondering

A jet-tail draws a long chalk line
To divide the silent cloudless sky
Would the jet succeed to race the sun
To arrive first in the evening

It is the supersonic Concorde
A mark of transportation techno
Its body hundred tons of metal
Flying same as birds size little
Its destination a point at another continent
Dwellings of people of varying intents
Whose lives changing in times multidecimals

Taking the Bus

Buses halt in front of the marked stops
Like ferocious animals sighting prey on the spot
Their huge bodies squat and scream
Panting their lungs rapidly let out foul steam

Toxic steams fill the air around
Spurting heat mixed with poison odour
Along various routes buses travel
Once in every few minutes they stop without fail

Passengers must wait no matter what
To inhale the toxic air before they board
Lucky is the one who arrives just in time for his bus
Most others have to keep watch and suffer

If by chance your bus takes too long to arrive
Taxises often come along to offer you a ride
They are like darting animals looking for small prey
They leave less pollutant as they run and stop in haste

People say the bus is more environmentally friendly
For all its polluting ills it carries passengers many
Try to stand a moment at a bus stop in the summer heat
You will not be a Smart Alex when you speak

Waiting

Me and my Mom ride
Up elevator flights
To arrive in a room people filled
Everyone ill

I watch the certificates on the wall
Wondering what it takes to get them all
Our ears are tuned sharp
To note the moment our turn is up

Entering the door we submit to the doctor's examination
With unconditional trust we respond to any question
Receiving the treatment the waiting is worth its time
The experience provides a peace of mind

Birth of a Poem

A poem is born
Against dark clouds
In a low grey sky
It hurls lightening bolts
Thundering crayons
Drawing at times a piercing chalk line
At other times frightening flash
To resolve in pouring
Life-giving rain

A poet coaxes the words
Lines that cry out meanings
Long trumpet calls or piccolo brittles
Drums summon and strings invite
The heartfelt emotions
Languid or harsh
Nerve shattering or soothing
Verses ready to foster
Unity of readers and the poet

Carefree

Just now I have nothing to do feeling free
Information flows attract response from me
The day is so beautiful serene at its end
Brush in hand I write to bid the sun adieu once again

My Teacher

You held my hand in my first day of school
You led me gently to sit on my assigned stool
You stood before our class tall and kind
You used words and pictures to expand our minds

You assured me I can joyfully learn and achieve
My obstacles and doubts you provided ways to release
My interests and curiosity you always encourage and applaud
To them you add the fun of venturing to explore new shores

You explain ideas and concepts simply without a preach
You use real life examples to connect with us each
You tell us stories about the wonders of oceans and eternity
You inculcate in us a sincere respect for parents and community

You share with us the unlimited powers of Nature and human wit
You show support and assurance whenever I feel inadequate
    ready to quit
You cheeringly urge me always to try hard aiming high
The hours and days we spent together have tempered my pride

Your cares and love have planted in us a love for country and humanity
You truly exemplify what the teaching profession is noble in reality
I am boundlessly grateful to you for your kindness and direction
You have laid a foundation for me to lead a life of commendation

My Life at Eighty-six

On this valley plain I walk and often loiter
Chatting with trees flowers bees and others
In the open space my body and soul wander together
Freed of gain or loss of matters fame or whatever
My pursuits and struggles are mostly left behind
Driven out from my inquiring and worrying mind

My strives are few but still real
Managed in varying paces and forces by will
The will for a happy and meaningful life
With caring for self and others combined

Scenes and lives in the universe I know
They live in connection and harmony as bestowed
With no competition trials losing nor winning
Just singing loving being and procreating
A squirrel is valuable as a bull
A drying pond is important as a river flow
All watched by a compassionate force from on high
Often left alone with appreciation in abide
Thus I live on satisfied and grateful hours and days
Enjoying peace and beauty my time runs on no haste

Insomnia

The muzzling days of childhood and youth
Interactions with stern parents tearful
Memories jostle now vague now vividly hold
In dreams where no dream is viable

The night sky is black like carbon paper
It prints what is marked in its fold
I repeat the prints lying awake on a desert pillow
Sands stretch moving hither and fro

My head is the interior of a mirror
Unbiased only ready to register
Whatever appears passes away and over
Blank in sleepless nights seem forever

I wish not to try pills purple or grey
Just give me a garden of roses their fragrance spray
A sense of peace in infinite ways
My calm mind resting in an open bay

IQ and EQ

If seeds and flowers could tickle the mind
Intelligence is on way to self-satisfy
If only computer games can help pass the time
Intelligence is on way to self-stifling

If you walk pass a dying bird without a tinge of feeling
Your emotion quotient is non-existing
If killings in war seen on TV are watched as you eat popcorn
Just forget improving your EQ scores on and on

Chinese Bell

Gigantic metal bells rung in open space to call people to play
In fields and village squares approaching end of day
Dances and delightful melodies celebrate bounty harvests
Those scenes appeared in China during many eons of years

Chinese bells are pear-shaped metal constructions
Their sounds resonated in open air in bright or dark visions
For years bell tolls signified the heavenly powers of ancient emperors
They summoned officials to court and sent edicts to rule commoners

Daoist and Buddhist temples have huge bells hanging in towers
They are rung to soothe the souls of living and deceased worshippers
On festive times bells toll continuously to enhance celebration
They advise people to reflect past deeds and set hopes for
    future aspiration

I often wonder what lyrical songs stars speak
And if mountains bellow messages for the strong and the weak
Torrents of mighty rivers carry moral themes in their flows
The ever shining moon murmurs silent messages to comfort our soul

Bell sounds are the sacred voice of the great Buddha
To help humans eliminate excessive desires and woeful powers
To treasure family joys and fond memories for lasting humanity
And to further human dignity wisdom and harmony

In our heritage each person has 108 annoyances per year
    needing clearance
Clanging the same number of notes promise to bring sublimity caring
    and defence
Buddhist monks use special tolls to help relieve sins of the deceased
And to rest the soul's drifting voyage in eternal wanders

Do listen to the bell melodies of the new year for calm and
    spiritual peace
And stand still before a temple bell to show respect and affinity
Bell chimes carry discreet messages for our activities in body and soul
O bell how your pristine voice direct us sitting or in travel

Section II

Nature We Belong and Behold

Dawn

A cloud on top of dark trees
White as the belly of a fish
The sky a pale blue
Half washed with greenish hue

I watch the elements from my balcony
Counting what else is in company
The cloud suddenly turns into pastel pink
To the white sky now a bird joins in

Not a moment the cloud is out of sight
The sky a pure mirror yellow and bright
Golden splashes wash the dark trees green
The universe serene a day begins

City Dawn

A single bird calls trees to wake
To announce the awaiting day-break
The sky a greenish grey
Tinted with a pastel yellow haze

The park is yet to be occupied
Up high a lone eagle leisurely glides
Suddenly a distant siren interrupts the quiet peace
A possible tragedy in precarious urban existence

An Eagle

It is a windy day
The sky a dark grey
Tree tops on sway
Up the air an eagle glides looking for prey

It soars high far and near
Sweeping at lightning speed with no fear
Suddenly it plunges down off my sight
I wish it has a catch it might

Each Day a New Spirit

Nature's sky seldom features a single colour
Nor does the sun come out in one manner
On this day's early hour I look
The sky lines the hill-tops with a yellow stroke
Above is this huge patch of charcoal grey
Further up is a greenish vast canvas on display

Where strips of light clouds streak gently by
Across a silhouette of tall buildings they extend wide
Haze on the horizon is testing its energy
To screen an enlarging orb its radiance not ordinary

I ask the neutral trees what's going to happen
Surprised they remain silent for the moment
When dawn appears its brilliance never a repeat
Each time it comes it brings a brand new spirit

Nature Cycle

A snow owl
Flying low
Covering distance
O'er a silent plain

Under frozen pebbles
Are found
Small bird bones
White as snow

Eagles Glide

Eagles glide in the air free
Above tree-tops they appear
From the ground they swiftly soar
Down they dive like free fall
I love to watch them my head held high
How they appear and disappear in the sky

Early Morning

Early in the morning
The weather is pleasing
Up in the air a lot is happening
White birds fly pass black birds glide
High and low and in circles they fly
How I wish to also try

Birds Sing

I get up in the morning before six
To prepare for school lots to fix
The hill outside my window is noisy
Birds sing loudly in a cacophony

On Sunday I wake up my Dad as usual
We listen to birds sing careful
Try to identify which bird is in solo

The loudest is a cuckoo
Tree sparrows busily chirp *jiu jiu*
Now and then calls the laughing thrushes
Accompanied by songs of yellow swifts

Dad says Hong Kong is a bird's paradise
Home or more migration stops for nearly 500 species
Attracting bird watches worldwide in all seasons
We are lucky to have such a rich variety
I now know the songs and habits of 87 birds friendly

In Praise of Autumn

A season you are
Not like any other
Golden leaves lingering on trees
Ready to mix with other colours to please

Wonders in your cloudless sky eternally shine
Bounty melons fruits and corn witness farmers' pride
Where sheep and cattle graze meadows are left bare
Migrating geese get busy their long flights prepare

A season of earnest hope and searching reflection
Soon winter will arrive with threats of repression
Maturity and harvest mark your riches
Together with delicious Thanksgiving turkey

How many people ascend to the platform of maturity
To experience the fruits of life with no worry
To enjoy what's present and forget acquisition flurry
To enter winter the season to be

Ah autumn you are a season of poetic fonts
With life events and struggles more than half gone
Duties and responsibilities behind happy and forlorn
Anticipation of renewing aspirations lifelong

I say autumn is an exhilarating season
Nature's enchantment triggering toils and tears for a reason
Urging our body and mind to access far-most reaches
Our tempered wishes feature creative strives and successes

Whence the limits of life appears in sight
Times before autumn should have prepared us right
No matter forwards or backwards we look
Satisfaction and pride will be vividly seen in our book

Clouds I Wonder

I wonder happy as a cloud
Drifting high or low so proud
White I am reflecting sunshine
Dark and heavy I rain day and night
Zen masters praise my courage in poem writing
Music masters sing my melodies and beauty
Artists paint me in fancy shapes and colours galore
Students scratch their heads when asked to describe me more

I wonder sadly on a dark cloudy day
My farming folks hurry to gather their drying hay
Villagers call each other to repair their leaky roofs
Philosophers pace the lonely lane to find proofs
The weatherman predicts rains probable
Grandma insists that I carry my raincoat
Could there be joy on a cloudy day
Farmers pray for rain dancing so gay

I envy clouds so freely roam
Appearing at will in shines and glooms
Beneath freezing clouds cattle happily graze
Before sunset clouds turn crimson orange and grey
Darkness reigns when clouds and sun not seen
Nature follows the beats of cosmic themes
Humanity is enriched with clouds floating in threesome

Floating clouds move in space and time so handsome

I wonder why clouds look like stones and marshmallows
They are wonders to bouncing children as to old-folks in solo
Could clouds hide behind the sky so shy
They could if heavenly father says not they are behind
People say every cloud has a silver lining
Could clouds prefer diamonds for clothing
Friends believe I am benign wondering how clouds might
It is for fun and beauty that I do so in hearty delight

This Hard Winter

The clock's minute-hand
Trembles slowly to land
On never-never land
Reading on I pretend
With no man-made heating
The cold pages are not turned
Under my seat hide my hands

The wind roars
Ruthless like the gnashes of a monster
Its teeth steal through naked space
Around windows and doors

Considering sleep and dreams
Warmth under blankets
Sandwiched with a thick mattress
They too wait in frigid cold
For human body heat
And summer dreams
The night shivers on
Future fairy-tales will tell
This winter is very cold

Weather Foretold

I remember Mother showed me the heavenly signs
That foretell the weather following from behind
We watched a clear sky up distant hills
Through a sea of golden rice all fields filled

It was in November in late evening
Before darkness enveloped everything
A white flash gave light to a dark grey sky
More appeared one after another low and high

Flashes in the east means a sunny day tomorrow
Flashes in the west will bring continuous rain
Flashes in the south heavy dews will follow
Flashes in the north strong winds will blow

We were seeing flashes in the north that evening
Everyone busily harvested the ripened rice in the morning
Gusts of northern winds kept the farmers reaping in haste
Else the bumper crop would be blown some grains fell to waste

Seventy years now from the day Mother sang that song
I have tested its predictive accuracy on and on
How I marvel the old wisdom its power to guide life
Even when no science awards help to augment the pride

In Gobi Desert

For years I fancied to scoop up warm fine sands at the Gobi with
my hands
And watched the sun dance as I let go the powders for breezes to fan
Camel bells chimed lonesome songs as the troop inched the vast terrain
on and on
Under the burning sun cool shadows cast on sand dunes short and long

My visits to the Gobi echo my fancies in varying scenes and feelings
Romantic thoughts of a vast hostile land embracing a tiny me is
exhilarating
Mongolian language names the Gobi as a very large and dry identity
Miles of sand dunes and rocky hills surround green oasis show
alluring beauty

I once visited the Gobi in a jeep travelling on gravel roads under a clear
blue sky
Along a ten kilometre way my driver friend Sarnai told me facts and
fables far and wide
On one spot rest fossils of Mesozoic dinosaurs which lived hundreds of
million years ago
Turning towards the Himalaya Mountain Sarnai stopped the car and
invited me to follow
Up on a red rock cliff we found symbolic messages ancient Buddhists left
for travellers

I asked in silence how people dwelled and preached here with sun moon
and whatever
At the camp that evening I realized that life and time in the Gobi were
very special
If one asked the stars the five Ws in human thought then all answers
were celestial

On my eightieth birthday Sarnai and I visited the Gobi savannah riding
on camels
For what awaited us to see and experience it was not an ordinary travel
Sitting on a seven feet tall camel my vista was far and wide
Feeling intensely emotional I stretched my arms to grasp earth and sky
White clouds play with snow peak mounts roaming merry
Down on the rocky outcrops and steppe plain solitary trees stand
like fairies

Hours later we saw a distant lake minoring the steaming air up high
A huge sand dune juxtapositions a field of grass and flowers
thriving wild
I climbed up the sharp edge of the sand dune to survey the scenery
in all directions
The most alluring view was the footprints on the soft sands attesting
my own action
How often could anyone review his travels so solitary on grounds
soft and rocky

His body and soul move in sublime actions toward an all possible destiny

Towards sundown I heard distant tolls of temple bells signalling
    the end of day
Sarnai waved to hurry me to see the hundreds of flowers special
    to the place
We strolled among needle grass and bridal grass blooming so profusely
We examined the blue and purple flowers on green stems of
    the desert lily
There were the Creosote bush foxtail agave and red pancakes
    with broad leaves
Among Joshua trees thrive Golden Barrel and Prickly Pear Cacti
    showing lusty beauty
They all stood in elegance in the majestic fiery of the sunset colours
A mighty eagle rode in the rhythm of winds with its wings
    of golden feathers

Thoughts Visiting the Glacier

Grand is the Arctic scenery
Sky and ocean-land in harmony
Billion-year snow sitting stationary

On helicopter I marvel the intricate beauties of the icy territory
On the ground my breathing lets out cold mists not ordinary

What wonders can Nature's hands create
Making immobile glacier its flows in wait
Alarmed are we to find fellow species extinct today

The cause is pollution induced global warming
Survival requires curtailment of desires for material wellbeing

By the Ocean Beach

I love to hear whispers of the ocean
Like drums in winter season
Rolling quietly in unison
To awake my mind's vision

Often I walk on the sandy beach
Touching a lone shell small as an inch
Orchestrating a dreamy melody
To narrate an intriguing story

O what wonders could oceans be
Their immense bodies and sensibility
Kindles what shells have to narrate
Oceans border humanity of all heritages and races
Diverse peoples believe and yearn for peace always
Regretfully they have found no way for true armistice

My Autumn

From my apartment tower on Floor Forty-Three
I watch the colourful extent of the Don Valley
No word could describe the autumn beauty that pull nerves high
Only mother nature has this power to enable human emotional drive

Inviting crisp air touches my face and nose so refreshingly
Beyond the greenish purple mounts the sky awaits the sunrise patiently
I ready myself for Nature's dramatic show as I do every morning
My whole being focuses on the unveiling

In an instance trails of ray shoot up the pearl sky shining with no bent
From the horizon to my presence a thousand pallets vividly present
Tens of thousands of trees aflame in multiple vibrant colours
They emit firing reds squash yellows burnt sienna water blues mild
    pinks whatever
Dark lines of tree trunks and branches sheepishly radiate
Like artists using their brushes to create magical shapes

I try to catch all the wonderful images vivid and remote
As they appear and change in the immense space to denote
The gland powerful imprints of our dear universe behold
It is a humbling but nourishing experience for me to hold

Sunset on the Don

By the balcony of my high-rise home on Floor Forty-three
I command a gland vista of the Don Valley
The pictures view changes every day and season
Revealing Nature's unlimited power in motion
The flow of the Don is veiled by shading trees
Birds trail the waters for fish revealing the river's course free

Inside the dense woods numerous animals and insects live a
    happy community
A walk on the valley is a body exercise and biological discovery
One could not but admire the earth's nourishing capacity
It is a path to better understand our universe and humanity
From deep antiquity to the present day mother earth has held
    up life dignity
For birds bees butterflies squirrels mice and colourful leaves in variety

The universe revolve in four cyclical seasons
Each has a promise and a willingness to shift for good reasons
Spring is for birth and development hopes and gaiety
Summer is for hard work dreams and creative responsibility
Autumn is for harvest enjoyment and captivating beauty
And winter is for repose and snow and a wait for spring surely

My favourite season is autumn ever since I was a child
Beautiful and nourishing images nudge my emotions fly

They precipitate warm memories day and night
To accompany my destination onward and bright
And a sunset in autumn has a special personal meaning in my life
It yields its glory gradually when the sky grandeur is high
It touches my soul up down left and right

I stand here awaiting to embrace Nature's magic drama
   in subdue beauty
And listen to the whisper of dancing leaves singing songs of piety
In time there will appear twinkling stars singing divine psalms
To welcome the moon's luminous charm
All these help me to review my long years of cares and neglects
   in providence
Overall I humbly say I am satisfied free tranquil and happy in essence

The moment arrives when the sky is ready to change looks of simplicity
Behold under fleeting dark clouds a blood red ball hangs motionlessly
The burning sun now shows its blazing glow before it says goodbye
Throughout the valley trees and earth subject themselves to a new dye
All is quiet to await a change of colours and lights
Then comes the final fall inducing people on watch to let go thrills
   and sighs

Cosmic happenings real or abstract that human intuition pertain
The wonders of human beings rest not only in beauties seen

They nest in satisfaction nobility and tranquillity that together make up
   happiness
Such state of being accepts personal faults and forgive injustices in
   destiny
I will do my best to achieve Buddhist karma in a world of confusion
   and irony

A sudden gust of wind reminds me that I am not wearing a mask for
   Covid-19 virus
I have left my Apple-13 uncharged to avoid receiving political lie
   news all hours
I stretch my arms to embrace the celestial gifts in this moment high up
   the Don Valley
They empower me to be a bearer of love to further our sustaining
   continuity

Algonquin Park
(Tune: Qin Garden Spring)

Late autumn in eastern Canada
A thousand hills in red and lavender
Ten thousand trees in brown and yellow splendour

Driving north from Toronto
On both sides of the roads
Leaves display brilliant colours fold after fold
The world is a show of fantasy fables

Gearing my car up full speed
Fields and valleys fly by like rushing steeds
In glee my mind serene my spirit upbeat

Under the searing sun
Blue and green mountain ranges quietly run
Faint clouds up high sky engaging in catching fun

Silver white lakes shine like mirrors
Reflecting a hundred tinted grasses vying for glamour

Checking in a cabin in the forest
I give my tired body a little rest
Breathing deep in the fragrant fresh air
My lungs inflate and my mind in flair

All worldly struggles result in loss or gain
They are but figments of shame or fame
These Nature's wonderful charms are mine to obtain

In high airy spirit
Holding a rod and wriggling baits in my hands
I fish in subdue pleasure till the day ends

On Air from Toronto to Hong Kong (Tune: A River in Red)

The silvery bird soars
Above all clouds float
Leisurely gallantly it goes

Lo the expansive sky
Serene luminous fine
My mind opens valley wide

Twenty-three days I visit places familiar
Forty-six years what ups and downs my career

I went abroad to seek knowledge to apply
My venture was not to keep wings from flights

Now that I had gone through the universe
I dedicate to promote world-wide my Chinese culture

What events in the old country
Deliberating current scenes and resent history

So much had changed for people
Their souls afflicted ample

How I watched cultural legends and lore destroyed
How I anticipate old values reaffirming with joy

Whence a delightful grand purview presents
New colours appear intense

A Fine Day

The elements tell
Everything will go well
On horizon a line divide
Skies and ocean colour lime

Ripples like fish scales silvery
They dance in harmony
In mountains near and far away
Birds rouse choruses in haste

Swaying trees move to please
Friendly cool free breezes
Heaven is now in glowing delight
Clouds stay humorously high

Colours frolic randomly with whites
Their images a magnificent sight
In time they fall
Down the mirroring ocean floor

Where I stand on moist sand
My feet cool kneaded by tender hands
Hesitant tides kiss the shore
Their bodies transform into fluffy froth

When I awake from awe
I see a world its spirits highly soar
A voice nigh announces quietly
A fine day will keep me company

Sunset Powers

The sun sets loud
Like old folks about
A day is like a life time
Events of joy and sorrow chime

Above the ocean loams a fire ball
It stands in luminous dignity tall
A constellation of colours silently speaks
Their glamour like lit cars on city streets

Yellow orange crimson burnt sienna white
Singing in trumpet roars far and wide
The lustrous orb falls on the lap of ocean
Ablaze with ever burning emotions

Radiant glows are love eternal
In flickering gleams life reaches final
A last touch of the magic brush
Paints the spheres with defying wishes

The sunset of life serene
Going leaving gone unseen
Behind in the mystic sea
Dawns wake eager to breathe

Niagara Falls

Torrent rush coming from the sky
The river gorge deep and wide
Rapids reach the escarpment suddenly fall
Freely forming foams and thick walls

Birds see the water falls as swirls boiling
People on ground hear them thundering
Is this the Niagara world famous
Or nature's show of motion and sound mighty

I go for the rainbow arching high
And nature's power through a window of my eyes
I visit nearby orchards and vineries
To sample fresh fruits and wines heavenly
Then appreciate plays of Shaw and Shakespeare
They depict human confidence and fears

On My Seashore

There was a time I adore
The seashore
I would sit for hours
Watching tides run like galloping horses

Froths fly high
Reflecting colours of the sunrise
In rainbow brilliance delight
Winds carry birds winging wild

Heavenly music high and low
Telling me to set life goals
Far as horizon behold
Their accompaniment makes life meaningful
Blazing the sunset
Time eternal

Over the Sea on Trees

Let's foam the sea together you and me
On the flame of sun burning free
Till the faint lights of twinkling stars
When twilight hurries dawn to start

Let time forget us together
Happiness and bliss measured not by hour
On wings of dreams we fly
O'er hills and islets standing high

We build a nest precious be
On thick leaves up roaming trees
In a world of eternal peace
Where gods and living beings adhere
Dreams are transformed into reality
By cosmic blessings and human ingenuity

Times and We

When hills beam and streams flow Nature happily show
When winds blow heavy and light my problems go
When grasshoppers sing in the meadow
Between blue sky and brown earth I stand in the middle
When birds twirl their colourful feathers extending
When heavenly melodies echo silver bells pealing

To challenges large or small I meet with open arms
To all lives large and small I respect and do no harm
When Covid-19 rampage communities worldwide
I take vaccines and wear masks following health guides
When time races or sits still as I do my daily care
My mind and soul keep calm however phones dare
When I am with friends and family I am happy and at ease
Let's laugh loudly in a universe cycling toward infinity

There will surely be a tomorrow
When pandemics and distancing vanish total
The new year of the fierce tiger is here to create
Global peace and continuing prosperity for humankind's sake
Let's look up with heads holding high
Rains and storms will be mellow with sunshine in blue sky
Human hopes are for real strives not gloomy dreams
We will meet any challenge however difficult they seem

Touch Nature

When was the last time you gazed at stars quietly at night
And asked why happiness in life was so hard to come by
When was the last time you sat alone below a towering pine
To listen to the silent songs of the forest their melodies chime
Have you ever climbed up a mountain peak before dawn to wait
    for sunrise
And marvelled the splendours of the sky changing in colours as time fly

How often you thought of walking with family in a meadow flowers
    growing wild
And did not do so because you or a family member was too busy
    to abide
Have you ever held an injured bird and nurse it to again fly
Or watch an eagle soaring so freely as winds lift its spreading wings
    to glide
If you have been afraid to hold a wiggling worm because it is slimy
If you did not ride a desert camel because it stands too tall and
    tempered badly
How you often wished to dive under the sea to see corals dance with
    playful fishes
Or regret having not steered a glider from a mountain top to view the
    scenic valleys

Your family celebrated your Fifty-second birthday cheering your success
    in business

Afterwards you reflected on how you had slaved for
    other-expected successes
You dreamed that night that you were a free man by nature to do what
    you wished
The same dream recurred again and again in your sleeps almost nightly
Finally you took off a week from work to get in touch with
    Nature happily
And record your misgivings and aspirations in this poem accordingly

Our Love

The gentle intermezzi fill the air day and night
You and I loiter in the serene garden our hearts unite
O'er the green land and grey sea the yellow moon is low
Towering mounts kiss the sky where clouds float

Ocean waves cress and ebb habitually show
The sun clasp the earth urging all lives to grow
We smile in silence stirring every cell of emotion
Not a word is needed to record our love immersion

Where Destiny Dwells

When our earth is a dew flying thru space moon and stars blight
Cosmic songs of heaven melodies ringing storms in flight
The clearest way into life is through the opaque wilderness
A thousand windows open revealing charms of the universe

Do love Nature and humans characterized by magnetic beauty
Embrace the mountain forests they harbour paths to our destiny

Poets Hear

Songs of birds at dawn
Whispers of trees hiss on
Pleasing chimes clank in air
Waters murmur human affairs

Drumming awakes senses
Recognizance bereaves
Time stands still
Space moves not at will

Imagination and dreams keep company
All awaking words and images
To orchestrate meanings in soul
Like clouds leisurely flow

Where truth kindness and beauty together sing
Verses of aspiration and wisdom forever ring

Horizon

(I)

Where the sea ends
What's the distance
Where the earth ends
What's the distance
They are the horizons

I ask grandma
She tells me to use my imagination
Grandpa joins in to say
The higher you go up hills to see
The further your horizon will be

I ask my teacher in school
She praises me to say
The more you ask the more you know
To ask questions is to learn
Asking is the best way to know

(II)

Distant yet near
Something always hopeful there
By amber sight
Heat-waves and tide

By flights of imagination
Crawls of techno-transportation
By the sun at dawn and night
The horizon appears bright

Evening

What are evenings for

For remembering the wonders of sunset
Warm glows kept the sky in subdue beauty

For listening to the whispers of leaves
Harmonious with the sounds of quiet peace

For awaiting the twilight of stars
Together with the moon's luminous calm

For reviewing the beginnings in life
And enjoy the fulfilling endings

For relaxing by a warm fire
Sparkles intensify the scent of pine

For anticipating a hard-earned repose
Thereafter the hopes of dawn

For simply enjoying a peace of mind

Listening to Beethoven

Dream-like melodies
Thunderous roars
Tranquil pauses
Expansive orchestration
Bloods swell deep in my heart
Shallows murmur in my ears
His are urgent insistent sounds
Music for the hungering soul
He wills to grasp infinity
A valiant show of Romanticism

My Sandy Rill

I stand again on my sandy rill
Like the years when I was here still
Watching its flow meanders at ease
Memories gradually release

No star can be seen in the blue sky
The altered landscape pale and wide
A moment of silence calls back sweetness
When the moon showered me with kindness

Sixty years had gone fast
So much sorrow had treaded pass
Now when I feel the cool water soothes
I erase those pains that I had gone through

My Sandy River

Meandering thru green paddy fields and olive trees
Wanders my sandy river so green and carefree
In her washed bed pink sands rest leisurely
Following her murmuring flow I wade and sing easily

Where she reaches a pond shrimps and fishes thrive
I spend my days trying to catch them avoiding the flies
And when summer brings in the burning sun
Me and my chumps find solace swimming along her run

As the day ends accompanied by melancholic music tunes
We run home carrying joy and catches not a moment too soon

Come Home Swallows

In spring
Swallows arrive
Around my eaves a pair fly
Could they be here last year
They now busy nesting
Baby birds are expecting
They grow so quickly
Learning to fly repeatedly
One day they disappeared
Gone and free
Will they return again next year
Home for sure

Returning Swallows

Swallows always return here in spring
They delight seeing flowers bloom along the lane
Gliding beneath the eave they flex their wings
Mud in beaks they build nests for the young to live in
Flying in pairs they contest in precision not speed
Once their young calls for feeding they toil indeed
Parents take turns bringing nourishment from fields and hills
Generations repeat the same love and care by Nature's will

Nature's Sigh

In among tall oaks I hear an owl cry
Its unidentifiable vowels enter my mind's eye
'Tis not music but Nature's sigh
The elements are stressed by excessive human desire

Where peach blossoms beautify the night
I walk in circles asking why
How people consumes easily on needs derived
When behaviours were not tempered by caring minds

Droughts bake the earth where growth was fine
Floods torment crops and trees their fruits denied
The poisoned waters no longer house fishes to survive
Where ill winds blow new diseases gladly thrive

Angels grieve o'er an ecosystem being spoiled
Dews tear to help farmers on their baked land toiled
How nature's elements sigh and weep
How their sops echo the owl's cries deep

Northern Lake

Up north near the Arctic circles
The lake is black and gold
Sprayed with colours red and yellow
Dazzling lights diamond white as snow

My oars shake up a world of cold
Hands stiff unable to hold
Hurrying fishes sing and echo
The spirit of blackness behold

Through lily skirting trees
Stars appear between scanty leaves
In quietude
To express silence of searching souls

Are you not deafened by such still sirens
Drowned by such beautiful gleams
The stars ask no questions
The answers nest in their blinking smiles

On Deck

I lie on deck drifting in the Pacific
In a tiny boat under a ceiling of stars terrific
Nobody around to speak to the moon hangs high
A gentle wind brings moist drops quickly pass by

I am grateful by the absence of nothing around
Free from annoying worries and noisy sounds
The crescent moon mute so I can dream
When awake in darkness she still dimly gleams

Progress the Price

Forests of factories
Standing luminance
Like beeping lights
On a plain people once lived in delight

Chimneys towering
Baked by fire and sun
Fuming out acid air
To no joy of eyes and nostrils

From the sewers flow
Waters red and purple
Draining into lakes and rivers
Where shrimps and fishes in toll

Green paddy fields are now cement grounds to behold
Hills flattened now crisscrossed with paved roads
Where trees loam high-rise dwellings have become the mold
People come and go with no community possible

Whence began and for what end
This beautiful land of old
Had become a symbol of progress
This place of my birth

Progress Oh Progress

How I yearn to see
Paddy fields shining green as should be
Clouds chasing the shadows they cast
Blue sky presenting beauty and pleasure that last

How I yearn to feel
Pines and flowers their fragrances the air filled
The moist cool earth at dawn under my bare feet
The juicy sweet of flesh cucumbers tasted in summer heat

Could it be that progress means childhood experience no more
But factory smokes and acid rain and skin sores
That children be deprived of nature's wonders in haste or days
And the sheer joy and affection of New Year's Day after much wait

Could it mean incessant purchases of things one needs because ads call
And that ecstasy can only come from a lucky win forgetting of losses
Could it mean that we care not to exercise our minds and morals
But exist to survive the day as if there is no tomorrow

Black Rain Prelude

At hours two seventeen
I woke to see the sky in a web of sheen
Thunderclaps hit in an instant
Shaking my building from no distance
Terrified I turned to look
Ten more series of thunders broke
They peel and bolt
Turning me breathless fears provoked
Nature is showing its wrath
The spheres in a whirl bath

In between clashes and flashes
I saw trees on the hill bent and heaved
As if to beg for mercy from heaven
Spellbound my heart rapidly pounced
To join cats and birds in tree bosom
Wondering if they were safe and sound

No voice amid deafening roars and clashes
Perhaps there were silent chatters
To comfort and pray
That the elements soon normally behave

But no Nature still has its way
In the morning when people wake
According to what the media say
We will be having a Black Rain Day

By the Sea

Desolate shores swell with whisper
Ever so gently
As not to disturb the dwelling shells
Nor to brick the seaweed spell

When winds blow carrying tides ashore
The sea widens the horizon shone
My ears din with cloying melody
My soul drenched in nymphet symphony

On the Sea Shore

Mounting cliffs their rocks explode
Whitened by rains and shines of seasons old
They stand like sighs without expectations or hope
Only as a part of Nature's spectacular look

The Shipwrecked Lady astride on the horizon
Her skirts display wings in bright vermilion
Her lips sweet and divine
Murmuring love to the sea and sky

Conches are delicious I know
They also make music loud and bold
They are the charms the sea hides
They surface only by nymphs we so admire

Hills Are Empty But Not Vacant

Grandpa reads me a line from a poem
Bird songs heard in hills vacant
He asks me how it could happen

I watch the green hill from my window at dawn
Here and there are heard chirping songs
But no birds seen in bushes short or long

Suddenly a small bird rises from amid tree leaves
A second bird follows to keep company
Ah the hill is empty but not vacant
Just birds are well hidden

Grandpa is fond of poetry
He teaches me the beauty of language
How thinking with language we enrich knowledge

Fishing at Night

A cold November night
Winds cut as sharpest knife
Whipping up waves high
To explode on the rocks
Like thundering shocks

Frozen fingers can grip no longer
Hooking unwilling baits a tough matter
Casting an act easier
A swing to the yonder
Eyes can but hear the water

Then is the waiting
Everything unmoving
Deep in the black water fishes may watch
How lingering stars shine
To keep the dark clouds bright

My heart jumped feeling the rod jerk
It was a strong wish keeping me alert
Blood circulation seemed slowing
May be this is all I kept saying
I had caught an experience

Glory of a New Day

Hours after midnight
High above dormant city-lights
Hangs a cobalt blue sky
No wind or sound
When people are in deep slumber
The whole world is mine to founder

The near hill and buildings a silhouette
Farther mountain ranges in my mindset
On this huge back sheet of non-glow
Three big clouds sit still white as snow
As I stand in silence to watch the show
They turn dark-grey ever as gradual

Now the sky blue becomes pure pastel
Light green mixed with greyish white
While the city enjoys its last sleep
The world is awaken whence darkness keeps
In the gentle shine of twilight
The new day begins in exhilarating beauty

The Heron

Ruffled by gentle breeze
Skimming o'er water gracefully
The heron wings across in no speed
Then down a tangle of thriving reeds

The day recklessly hot
On my skiff I sit as I ought
Hearing fishes not seen
Small waves slap to sing

Nothing is moving
Where I search beyond the shore
Against the glare of water shining
The heron is no more

Leaf Boats

Leaf boats sail in wind
Following river rapids and flow
To rest in puddle ponds
Glittering like a dream born

A Rainy Day

Driving rain beats on the window pane
Winds chill blustering
Safe in my apartment with music divine
I attend to my mind's eyes

Green hills cheering clouds high
Overlooking a valley that thrives
Paddy fields of shooting rice
Their beauty bursting in sunshine

Wonders of Dawn

Beneath grey clouds and pale-blue sky
Scattering patches of orange pink rise
Further up a protruding cloud is silver-lined
To complete this fascinating picture of dawn

Hope is the mood
As people on the street jog and run for good
Every individual has a style of exercise
All share a common goal of health enterprise

In a moment
Like twenty seconds
The colours burnt
A clear bright day is born

Bright Moon and Buddha

My mother told me when I was nine
Life is fair and fine within Buddha's sight
You must be diligent and you must try
You are your own Buddha in life

The moon is the Buddha I learned at night
She always shines to keep the dark world bright
To give those who seek wisdom and peace of mind
A renewing respite after continuing strive

Quietly she gluids
To give all followers directions and delight
She is always right
Because in you she resides

My mother is now gone
But the wisdom she had left behind
Is forever in me shone my guide
She is my Buddha and moon lifelong

Tadpoles and Frogs

The village pond is shining green
Grasses adorn the banks a serene scene
A huge patch of dark-grey matters is seen
Tadpoles flourish in late spring
They waddle thin tails to move large heads
To show a mass of active aquatics

I visit the pond for interest weeks later
No tadpoles are seen anywhere
From amid the grass *guo guo* songs sung one after another
They are frogs grown from tadpoles I discover
They feature strong bodies and long mighty legs
They jump on land easily as swim in water

As Nature's grand gift they possess another character
When cooked with garlic oil they are a heavenly savour
You will surely have them in the finest gourmet dinner
How lucky we humans are in the cosmic sphere

My Forest Greens

Every morning at dawn
I watch my forest on and on
Trees number in hundreds
Their colours vary with lighting spreads

Greens in brightness and hues varying
All kinds of life activities moving and relating
How a forest harbours mysteries
Paths and bushes record past and resent history

My interest is in the changing greens
Dull grey vivid pine olive lime and lotus scenes
On black tree trunks emerald and neon leaves fly
Like butterflies dancing to decorate the placid sky

I once counted shades and hues of forty three
By winter some leaves fall the green numbers remain steady
On planetary perspective greens keep earth thrive and peaceful
Privately watching greens helps human eyes healthy and restful
Forests form the lung of our huge planet in the universe
Holding high the Green Peace Movement will help us to survive

Snow Fires and People

Is Nature giving or showing her wrath no one knows
In one day two feet of snow poured down to Toronto
Watching by the window of my high-rise apartment with a sound roof
I appreciate the amazing scene of the city feeling grateful
Against the huge deep blue shining sky no eagles fly
Along the ten mile extend of the Don Valley the world is pure white
Thousands of trees bow low to show their submission to Nature's power
Overwhelmed by snow their crowns show not their usual tower
In the quietude I hear continuous whispers brought by zephyrs
They express sympathy for the hungry birds and animals beleaguered

On the other side of our planet there was an under-water
    volcanic eruption
Part of Tonga Island sunk submerged into the Pacific Ocean
People float on stormy waves their houses and lands lost to
    cosmic exertion
The survivors need foods drinking water and medicine urgent
Supplies and rescue actions pledged are ever so slow in motion
How prosperous countries relate to their own conscience
Knowing their wars are responsible for these tragic commotions

Advanced nations have raised wars in faraway places for gains
They believe they have the right to abuse natural resources leaving ruins

Behold calamities have turned back to harm the same nations

Half a million forest fires had burnt out
   American mountains
The unceasing blazes killed all kinds of lives on more than five million
   acres of lands
Trees and animals eliminated will cause definite ecological imbalance
People ask why the most armed nation cannot put out wildfires
   in a hurry
And how leaders sit at ease seeing half a million citizens dead in the
   pandemic so eerie
Statistics on life losses appeared unsystematically and scanty
They emphasize on insurance payments most vividly

Poor forest management and global warming are blamed
   for the tragic fires
Plans for using advanced technology to prevent future fires are scanty
No report is seen on how to revive trees and animals to rebuild
   a healthy ecology
Might it be high time that we review our cares for humanity
Not only on how to keep mother earth healthy and continuous
But also on how to affirm the value of lives and human dignity

April Morning

Early morning
Trees shining
They stand tall and about
Over hills in and out
I counted forty-three greens
With flowers they gleam
Birds compete to sing
Tenor and soprano notes ring
The sky in pale grey
Tinted with suggestive pinks in haze
In an instant
The world turns white and crimson
A moment of strange silence
Even birds pause to listen
There on tree-tops appear
In a sea of colours the sky cheers
'Tis an April morning

Technology and Nature

From up a tree I look out far to see the horizon
Between me and the distance lies a gleaming ocean
Loaming high above hovers a cloudless sky

I am at a small private beach near Montego Bay
'Tis winter and the neighbouring cottages are not engaged
Serenity reigns when no wind nor breeze wave
Seaweeds washed ashore their fresh colours on white sand display

Suddenly as if by miracle a big white bird appears
Its majesty and angelic grace challenge the imagination
Could such a huge piece of heavy metal fly so slowly so light
On air suspended only by human ingenuity and insight

My heart is stirred
I am moved to pray
And to wish the travellers a safe journey
Wherever their destination may be

There is no question if technology can be
A part of natural beauty
In its grand constellation of appearances
Beauty is not in our vision but in our consciousness

Section III

Children and Parents

Good Habits

Good habits we keep
Happily we live
ABC
Learn to wear clothes
DEF
Learn to wear pants

Ten Little Fellows

Ten little fellows
Hands hold
To form a circle
We turn left
We turn right
We sing and turn many rounds
Ten times each side
Twenty times for both sides
We are all tired
Before we finish
We yell out loud as we wish
We are healthy and happy

Me Growing Up

When I was a child I love to play hide and seek
In my teens I wonder why grown-ups are so fond of money and meat
At university I admire a professor whose knowledge is deep and wide
Befriending me he confines that he questions and thinks timid and wild
As I turn to take up business for wealth in life
I became a slave working busily day and night

When I was a child I love candies and things sweet
In my teens I wonder why girls are hard to meet
I dropped out of university my parents showed understanding
For three years I drifted from odd jobs to dish washing
I met this caring kind old man who listened to my story
He advised that life is about striving to achieve and be happy

When I was a child I love to be with many other kids
At school we learn and play for days and weeks
In university I discovered that friends are more important than
    knowledge
Then I got married and wife and I brought up a family
I worked hard contributing to society and earned my dignity
Friends and family care and love in our harmonious community

On Swing

Here I go on a swing
A happy child of spring
Upward to see the sunny sky
Downward to see grounds green
Suspended on a strong iron beam
Centre of the world I am in
To my ears rhythmic winds ring
Loud and free I let out screams

My friend John comes to join the cheers
He wields up on air clasping the strings
Then we both ride on the same swing
Laughing going up and down and singing
Our fun increases as my Dad comes to see us
He gives us hot-dogs and cool drinks so marvellous

Together We Sing

We sit in rows
In unison our singings go
To praise beautiful hills and rills
Laughing and laughing
Our plains produce plentiful
Happy and satisfying
We sing to echo the roaring ocean
They connect us to all nations
With love and understanding
Our human family will thrive everlasting
On and on we sing

My Body and Me

I have ten fingers
They make wonders
I have ten toes
To far distances I travel
My mind loves beauty and novelty
It enables me to question and know
My heart feels and sympathies
It urges me to care and be friendly
So perfect are things in life
I expect nothing more from the sky

Me and My Shadow

Mom and I walk on street at night hands joined
Street lamps cast shadows follow us on
Sometimes in front sometimes behind
Sometimes on left sometimes right

Shadows appear one or two
Or nothing to follow
Shadows are strange but lots of fun
They stay with me in life all or none

I Am a Daffodil

I wonder and wonder
If I can be a flower
What should I be
A rose or cherry
Tulip or lily
A daffodil I will be

Standing proud
In yellow crowd
Over lake and valley
My gold colour is pure beauty

People visit me
Breath my scent free
Admire my tranquillity
And drink me as tea

Little White Rabbit

Little white rabbit
He likes to hop indeed
Eyes red and ears big
Furs silky his tail wiggles quick
I give him a carrot to eat
He munches and steps his feet

Number Song

One two three
I am free
Four five six
I can fix
Six four five
I love to write
Tree one two
I sing also
I am a happy individual

Dreams

I was flying amid stars in my dream
And down on hills and valleys in gleams

I am master of free actions
Roaming over oceans and mountains
Until I fell to a hay stack
Though no hurts I felt being packed

I cried out for a star to lift me up
My loud call woke me instead
Back to reality it is morning
Time to school studying

Dreams give me moments of being totally free
They fill my mind with fond memories

My Happy Dream

I cuddle my Teddy Bear tightly to bed
Hoping for a night of refreshing rest
Not before long
I fly on wings
Journeying over hills embracing a thriving plain

I see folks working in rice fields
High rise builders greet me to say hello
A nurse helping an old man to get sunshine
Children rushing to school hands holding tight
They all busy in useful strives

The sun rises to herald in a new day hopeful
I wake up realizing I had a dream delightful

Cat and Mouse

I have a cat to catch mouse
She *miao* and *miao*
The mouse runs out
Can you guess what about?
My guess is
Seeing a cat around the house
The mouse quickly got out
My cat's duty becomes null
She catches a ball as if it is a mouse
For fun and responsibility
What a wise cat she is
I clap to praise her heartily

Little Rooster and Big Red Hat

Little rooster
Polite always he appears
Saying hello to the morning sun
Father sun laughs happily ho ho ho
Giving him a big red hat to behold

Little Rooster Crowned

The young rooster is diligent
He is also responsible
Calling ook-ook-oook from the barn top
He wakes up people at dawn to work

Walking about in early morn
He meets Grandpa sun to say hello
Grandpa rewards him with a bright red bonnet
He holds up his head to walk like a hero

Mother Hen and Her Chicks

Mother hen rises early to *guo* aloud
She leads her chicks to walk about
Finding a centipede she happily shout
Come together let's all eat it out

Siblings

Siblings are heavenly gifts eternal
Brothers and sisters grow up in same burrow
Bonding is both easy and difficult
Learning to care love compete and fulfil

Siblings hold hands to discover and succeed
Drying tears whene'er one falls and bleeds
Siblings share joys and secrets openly and in discrete
When they play hide and seek they test the use of space for benefit

Born from the same mother siblings share similar genes
Parents love and teach them the values of life and spiritual hints
Parents show their kids practices of mercy and charity
And expect them to live a quality life champion of humanity

When siblings fight they learn to use their physical powers
Playing hide and seek they test space and blockage for hours
Sibling families produce cousins sharing kinship
In time clan members form communities and partnership

When we honour the golden ethical edits as follows
An older brother should be friendly and helpful
The younger brother should be humble obedient and respectful
Our social order will be in peace eternal

Songs

When I was little I love to listen to songs
Sung by my mother or anyone who came along
I knew not tunes or poets' thoughts
Nothing was more pleasing than words gently brought

The moon shines the moon bright
Her beams and gleams light up not only the sky
But all else that is in the universe
And the little boy whose heart nurtures harmony

The moon shines till New Year's Eve
She continues her brightness again the next eve
Where songs go the moon has more stories to tell
Followed by comets carrying omens occasionally fell

As I grow old I find songs no longer free
Not for love nor for dreams to be

Poets still sing with heart-rending airs
What I hear is not the moon songs I so much care

Speaking to Stars

I remember those summer nights
When houses were too hot inside
Family and neighbours gather outside
To rest on the pavement where grains got dried

Those were happy hours for kids
As we lied on cool bamboo mats heads on pillows of seeds
Comfy on our backs we watched out for the first star
Then next and hundreds near and far

To no avail we count them all
But to learn from old folks and grandmothers
The place and make up of the Big Dipper
The Northern Star and story of the Weaver

We hear deeds and courage of ocean venturers
We sing and whisper to show pleasure
Until the sky appearing pale blue and wide
To signal in the moon's rise

It is time we work out our wishes
To tell our favourite star in the distance
To help make us wise and healthy
To be happy and exemplarity

I recall vividly in memory
That I ask my star to set me free
To run through fields and up on trees
To scale the mountain and across rivers
To go anywhere I please
And to care my folks for years

Stars I Ask

September in mid-autumn the air is fresh and clear
My family goes up hill to watch stars near
Before the moon appears

Dad teaches me how to locate the Big Dipper
Mom narrates the story of The Cowherd and the Weaver
Grandpa says each of us has our own star
To help us develop and achieve near and far

I lie comfortably on a soft carpet of grass
And ask how many stars there are
Dad says no one in history could say definitely
Even galaxies number in thousands
And each galaxy contains stars in billions

I ask why are stars so many
My parents say Grandpa could tell reasonably
He says it is natural the universe has many unknowns
So we can explore and fantasize using our capabilities
It all makes life fun and challenging

In a moment the moon rises to shine overall
Stars recede behind the deep blue sky seen no more
Until the following night they reappear for sure

Our lives follow a similar order as the stars in universe
Now we dance and sing with vitality on the cosmic stage
Now we rest backstage enjoying satisfaction and peace

Stars and My Age

When I was one
My life has just begun
Now that I am six
I wonder what will come next
Of course age seven

In school I learn reading and arithmetic
At night I watch the stars numbering big
Sitting with Mom feeling great
I wonder there is a star like me turning age eight
A voice tells me eight million years is more like it

Twinkling Stars

Twinkle little star
Diamond bright you are
Eyes in the sky high and far
One in the west
One in the east
Thousands and thousands fill the galaxy

We Are We

I was five my sister three
We rode the wooden horse at ease
Front I was back was she
We skip and trot laughing giggly

When I went to school my sister was free
She cried periodically missing me
Free is not easy
Togetherness let us be

We will grow up like all siblings
Time to separate and develop differently
But we will always be
Sister and brother we are we

Jumps and Falls

I watch at the park a girl of five or four
Hands outstretched she jumps repeatedly to catch a hanging ball
The autumn morning sun is shining ever so soft
It warms and encourages the girl to jump more and more
Sitting near her mother is concerned her daughter may fall
She calls her to stop the jumps once for all
Nearby an old man mutters as the mother tells the girl to stop her fun
    was ample
'Tis not the ball but the reaching the old man mumbles

Good Vegetarians

Rabbit lovely rabbit
Vegetable and grass she eats
She loves to graze with cows and sheep
Seeing an eagle she runs to hide at top speed
Her fur is soft and white I love her indeed

The Circle

Grandpa helps me to draw a circle with a compass
The circle has a very special character
When you trace it from a beginning point to go around
That point will also be the end of your destination
The trace or journey is called a cycle

Grandpa says our life goes through cycles small or large
We take many actions for fun or purposes
Hoping to reach desirable destinations
To enrich life and to enjoy satisfaction

Crow and Sparrow

An old crow
Ah Ah it squalls
People everywhere scold
Little sparrow Gi Gi it tweets
I love to watch it hops and skips

Time and Distance

How far does the sea go
Where the sun sets low
Humming of whales across oceans
On air birds glide their bodies float

In my mind a voice says
Affection between you and me aglow
Together we enjoy growing old
As cosmic time holds

Gathering Mushrooms

Rabbit black and rabbit white
Gathering mushrooms on hills high
Little monkey and small deer
Joining in to help and to care
Rabbit and monkey
Deer and rabbit
Happily they gather mushrooms to eat

Mushroom Hunt So Fun

We chumps set out early at dawn to hunt
For wild mushrooms before the rising sun
The moist cool earth keeps our bare feet on run
A fond feeling remembered in years to come

I got one
One of us shouts in delight
As he sees a single mushroom on a hill side
Sure as a button more mushrooms stand on higher grounds
We pick and cheer as we gather them in rounds

Those were wonderful childhood days in rustic life
Where green hills and clear rills surround fields of rice
Where trees shade busy farmers under the burning sun
As people grow and harvest in cycles enjoying the fun

Friendship lasts to keep our spirits high
Gifts of nature are shared with gratitude and pride
Mushrooms thrive after heavy dews where cows had left their daunts
They are nourishing and delicious when cooked well-done

Butterfly

Flapping fluttering you butterfly
Amid flowers you merrily fly
You wear colours blighter than flower beauty
You fly high and low so freely
No wind could change your destiny

Butterfly butterfly l ask you
Would you alight on my palm momentarily
To let me touch your wings gently
And feel your powerful might
I will be so very happy in life

Be Good in Life

With what colour and language you dream
In what winds and gardens your thoughts beam
Accept and appreciate people their virtue seen or unseen
Accommodate and value people whose fault is not seen

Help and be charitable to those who are unable and weak
Care for parents and elders their love so devoting and meek
Love and be loyal to friends and country you belong
Your satisfaction and happiness ride on time streams drifting on

That Memorable Summer

That was a summer to be remembered
I waded thru waist-deep water along a river
My world was open full of splendour
Friendly birds I knew swooped pass over
Unseen frogs croaked under shadows of trees
On purple grass tops grasshoppers sing songs of mystery
A grey hawk wheeled to kiss the puffy clouds drifting
While I drank the wild perfumes of sweet winds blowing

Flashes on the horizon signalled rains to come
I dashed up the nearest tree seeking shelter and fun
My neighbour buddy was already there hiding
We screamed as the storm struck

Through the leaves we saw the fields on steam and mist
Between sky and earth there floated a thick layer of haze
In a sudden the bright sun shone again hot as ever
My friend and I took off our clothes to plunge into a nearby river
Freely we swam and jumped to cheer Nature's change
The kind of shift only a summer day could bring

On the evening things were calm and heavenly
Feeling drowsy my eyes saw bowing trees wavily
Cool distant hills glimmered feebly facing a widening plain
I fell on my soft pillow and the day passed free

Carrying me home my Mum's embrace is warm and neat
As always Mum could choose the right moment to teach
That life is for striving and care
There would surely be failures and storms here and there
After Nature's heavy storms sunshine would again fill the air
Thus my memorable summer had not only fun and daring
With my friends I grew up healthily to work hard and be caring

Children Then and Now

How I enjoy wading through the hillside brook to find fish
Sounds in the valley seem to answer my wish
I know not images of beauty poets find in Nature yet
Just jump with joy when a tiny fish is caught in my net

Murmurs heard from the brook says something to me
Distant hills and trees show time permanence and peace
If by chance a sudden downpour wet me all over
The cleanliness felt goes well with the cool delight ever

When my Mom's call for home resounds on all sides
It is time to run home fast before hearing her sighs
Happiness is a day spent with a good catch
And how a warm hug welcomes me with a smile to match

I now watch my grandson spending his hours and days in school
A three-year-old he attends classes to learn to count and to spell
To line up for turns to sing and sit on a designated stool
And to play computer games for fun and everything else

His parents enrolled him to school when he was eighteen months old
Least when he reaches two he be denied admission to an
    international school
To prepare him for interviews his mother drills him to smile always
And to answer questions readily

As he advances from playgroups to kindergarten next year
His teacher will discipline him like a soldier with no freedom to fear
I dare not guess what kind of a child he becomes reaching six or seven
He will certainly not catch fishes nor listen to songs from heaven

Section IV

Poems by English Poets
with Chinese Rendition

Invictus

—— William E. Henley

Out of the night that covers me,
Black as the pit from pole to pole,
I thank whatever gods may be
For my unconquerable soul.

In the fell clutch of circumstance
I have not winced nor cried aloud.
Under the bludgeonings of chance
My head is bloody, but unbowed.

Beyond this place of wrath and tears
Looms but the Horror of the shade,
And yet the menace of the years
Finds and shall find me, unafraid.

It matters not how strait the gate,
How charged with punishments the scroll,
I am the master of my fate,
I am the captain of my soul.

不敗的心

—— 威廉・亨利

受羈困漫漫長夜

那陰陽極間的漆黑深淵

我感謝上蒼神明

賜給我不可戰勝的靈魂

身陷生死關頭

我從不畏縮或喊嚷

縱受困於無情控制

我血破的頭不曾俯屈

憤怒與哭泣以外

瀰漫着地獄般的恐怖

不過面對無盡威迫

我堅持無懼

不管牢閘如何緊閉

不管罪狀何其殘苛

我是命運的主宰

我是靈魂的舵手

Bright Star

—— John Keats

Bright star, would I were stedfast as thou art—
Not in lone splendour hung aloft the night
And watching, with eternal lids apart,
Like nature's patient, sleepless Eremite,
The moving waters at their priest-like task
Of pure ablution round earth's human shores,
Or gazing on the new soft-fallen mask
Of snow upon the mountains and the moors—
No—yet still steadfast, still unchangeable,
Pillow'd upon my fair love's ripening breast,
To feel for ever its soft fall and swell,
Awake forever in a sweet unrest,
Still, still to hear her tender-taken breath,
And so live ever—or else swoon to death.

明星
—— 約翰‧濟慈

明星我願學你一樣堅定

但不學你長夜孤懸中天獨明

睜着一雙永不閉的眼睛

像忍耐不眠的隱士

或傳教士執看河水奔流

用聖水滌淨人世海岸

或守望輕飄下降的初雪

為山峯和原野載上柔軟的面幕

啊我不願但仍然堅定自守

緊貼着愛人的成熟酥胸

在甜蜜中享受柔和起伏

永恒清醒地受着甜蜜的騷動

我必要必要傾聽她溫柔的呼吸

如是永生或在昏醉中死亡

The Human Seasons

—— John Keats

Four Seasons fill the measure of the year;
There are four seasons in the mind of man:
He has his lusty Spring, when fancy clear
Takes in all beauty with an easy span:
He has his Summer, when luxuriously
Spring's honied cud of youthful thought he loves
To ruminate, and by such dreaming high
Is nearest unto heaven: quiet coves
His soul has in its Autumn, when his wings
He furleth close; contented so to look
On mists in idleness—to let fair things
Pass by unheeded as a threshold brook.
He has his Winter too of pale misfeature,
Or else he would forego his mortal nature.

｜人的季節
── 約翰・濟慈

一年有春夏秋冬定位

人的心靈同樣活在四季

他有生氣蓬勃的春天

當天真幻想迎來美好的一切

他有火熱的盛夏

叫他追想初春的甜蜜年華

用志高情深的夢想走近天堂

他沉靜的心靈染上秋天的風味

當他把翅膀和衣衫一同收起

滿足悠閒看世界的朦朧景象

容許一切繁華盛事

像門前的溪水在不經意中流過

他還有冬天和蒼白變形的臉孔

他不能超越人的本性

| **When I Have Fears That I May Cease to Be**

—— John Keats

When I have fears that I may cease to be
Before my pen has gleaned my teeming brain,
Before high-pilèd books, in charactery,
Hold like rich garners the full-ripened grain;
When I behold, upon the night's starred face,
Huge cloudy symbols of a high romance,
And think that I may never live to trace,
Their shadows, with the magic hand of chance;
And when I feel, fair creature of an hour,
That I shall never look upon thee more,
Never have relish in the faery power
Of unreflecting love: —then on the shore
Of the wide world I stand alone, and think
Till love and fame to nothingness do sink.

| 當我恐怕自己的生命終止
—— 約翰・濟慈

當我恐怕自己的生命終止

而我還來不及寫下我的澎拜思潮

還未讀完堆如山高的典籍

豐收那成熟的金黃穀子

當我在深夜仰望繁星

馳想雲層中隱藏的愛情故事

深感即使幸獲機緣之助

自己亦永無雲彩庇護

當我感到絕世佳人的呼喚

亦永不敢對她凝視

沒有仙人的權能

享受不渝愛情的滋味

我獨立岸邊沉思

悟見情與譽的虛無

| **On Death**

—— John Keats

Can death be sleep, when life is but a dream.
And scenes of bliss pass as a phantom by?
The transient pleasures as a vision seem,
And yet we think the greatest pain's to die.

How strange it is that man on earth should roam,
And lead a life of woe, but not forsake
His rugged path; nor dare he view alone
His future doom which is but to awake.

| 死亡

—— 約翰·濟慈

當生命是夢死亡可能是睡眠嗎
幸福的歡樂可是幻影的過去
瞬間的愉悅似是過眼雲煙
我們卻確認死亡是最大的痛苦

人生在世流浪是一樁奇事
要度過一些悲慘也不願拋棄
一路坎坷也不敢靜自思量
將來的死亡只是一種醒覺

| Daisy's Song

—— John Keats

The sun, with his great eye,
Sees not so much as I;
And the moon, all silver-proud,
Might as well be in a cloud.

And O the spring—the spring!
I lead the life of a king!
Couch'd in the teeming grass,
I spy each pretty lass.

I look where no one dares,
And I stare where no one stares,
And when the night is nigh,
Lambs bleat my lullaby.

雛菊之歌
—— 約翰·濟慈

太陽張開大眼睛
不如我看得清
月亮的銀色驕傲光輝
只能住在雲端

春天啊春天
我的生活好比國王
我躺在無邊的草坪
偷窺每一位美麗姑娘
我看無人敢看的地方
凝望無人凝望的景觀
等到黑夜臨近
羊兒會給我唱催眠曲

On the Sea

—— John Keats

It keeps eternal whisperings around
Desolate shores, and with its mighty swell
Gluts twice ten thousand caverns, till the spell
Of Hecate leaves them their old shadowy sound.
Often 'tis in such gentle temper found,
That scarcely will the very smallest shell
Be moved for days from whence it sometime fell,
When last the winds of heaven were unbound.
Oh ye! Who have your eye-balls vexed and tired,
Feast them upon the wideness of the Sea;
Oh ye! Whose ears are dinned with uproar rude,
Or fed too much with cloying melody,—
Sit ye near some old cavern's mouth, and brood
Until ye start, as if the sea-nymphs quir'd!

| 詠海
—— 約翰・濟慈

你收藏着永恆的絮語

滔天大浪湧拍荒涼孤岸

何止千岩萬穴

留着女神郝卡忒咒語的回音

你時常和順安詳

收留最小的貝殼

數天不許移動

天風驟降

當你眼睛受惑倦慵

請張開飽看大海的汪洋吧

當你的耳朵受震欠聰

或聽膩了重複的韻調

請閒坐在古洞面前吧

冥想傾聽美人魚的歌唱

O Solitude!
If I Must with Thee Dwell

—— John Keats

O Solitude! If I must with thee dwell,
Let it not be among the jumbled heap
Of murky buildings: climb with me the steep,—
Nature's observatory—whence the dell,
In flowery slopes, its river's crystal swell,
May seem a span; let me thy vigils keep
'Mongst boughs pavilion'd, where the deer's swift leap
Startles the wild bee from the foxglove bell.
But though I'll gladly trace these scenes with thee,
Yet the sweet converse of an innocent mind,
Whose words are images of thoughts refin'd,
Is my soul's pleasure; and it sure must be
Almost the highest bliss of human-kind,
When to thy haunts two kindred spirits flee.

假如我必要與孤獨同住
—— 約翰・濟慈

哦孤獨假如我必要與你同住
但願不相處在灰堆砌起的大廈
請與我同攀峯頂大自然的望台
看近在咫尺的山谷
綠茵繁花遍野中那晶亮河水
讓我們靜守在枝葉蔭蔽中的幽亭
看闖鹿驚起仙人鐘花叢裏的野蜂
雖然我情願伴你尋訪這些美景
同純潔心靈親切交往
傾聽他心中的精言妙語
實是我心底的至美樂事
亦是人類的最大福氣
尋着投契心靈你我一同奔馳

To the Moon

—— Percy Bysshe Shelley

Art thou pale for weariness
Of climbing heaven and gazing on the earth,
Wandering companionless
Among the stars that have a different birth,
And ever changing, like a joyless eye
That finds no object worth its constancy?

問月
—— 雪萊

你可是因憂心而面色蒼白
莫不是為了凝視大地夜夜登天
永遠孤身漂泊
在身份不同的眾星之中盈虧循環不輟
像一隻欠缺歡樂的眼睛
找不着對象共同長久眷戀

The Lake Isle of Innisfree

—— W. B. Yeats

I will arise and go now, and go to Innisfree,
And a small cabin build there, of clay and wattles made;
Nine bean-rows will I have there, a hive for the honey-bee,
And live alone in the bee-loud glade.

And I shall have some peace there, for peace comes dropping slow,
Dropping from the veils of the morning to where the cricket sings;
There midnight's all a glimmer, and noon a purple glow,
And evening full of the linnet's wings.

I will arise and go now, for always night and day
I hear lake water lapping with low sounds by the shore;
While I stand on the roadway, or on the pavements grey,
I hear it in the deep heart's core.

因尼斯夫莉湖島
—— 葉慈

我即此動身到達因尼斯夫莉
就地蓋建一間土木砌造的小築
躬耕九坑長豆及蓄一窩蜜蜂
逍遙獨居在嗡嚷的林地

我將享受來之緩慢的和平心境
靜聽湖潮拍岸的清音
子夜朦朧正午紫光
藹暮光中充滿朱雀翻飛的趐膀

我即此動身前去那不分日夜
都聽見湖潮拍岸的清音
不論行走大路或駐足灰石路旁
潮聲都在我心底迴蕩。

Symbols

—— W. R. Yeats

A storm beaten old watch-tower,
A blind hermit rings the hour.
All-destroying sword-blade still
Carried by the wandering fool.
Gold-sewn silk on the sword-blade,
Beauty and fool together laid.

| 象徵
—— 葉慈

一座風雨摧殘的古瞭望塔

由一位盲的隱士敲鐘報時

一把所向無敵的利劍

掛在無知流浪者身上

金色織錦包着劍鋒

美人與愚人同平放在一起

| The Coming of Wisdom with Time

—— W. R. Yeats

Though leaves are many, the root is one;
Through all the lying days of my youth
I swayed my leaves and flowers in the sun;
Now I may wither into the truth.

智慧依時出現
—— 葉慈

樹葉茂盛源出根原
那些青春虛妄歲月
我曾在陽光下玩弄繁花
如今且讓我抱緊真理

The Fisherman

—— W. R. Yeats

Although I can see him still—	雖然我仍認得他
The freckled man who goes	那滿臉雀斑的他
To a gray place on a hill	習慣前往迷濛山地
In gray Connemara clothes	身穿康瑪拉粗衣
At dawn to cast his flies—	在晨曦中揮竿釣魚
It's long since I began	如今過了一段時間
To call up to the eyes	當我記起
This wise and simple man.	這樸實智者
All day I'd looked in the face	我整天向那張臉凝視
What I had hoped it would be	尋找可能的寫作信息
To write for my own race	敘述我們民族與現實
And the reality:	
The living men that I hate,	那些我憎惡的活人
The dead man that I loved,	那些我愛的死人
The craven man in his seat,	那些掌權的懦夫
The insolent unreproved—	那些逍遙法外的無恥人士
And no knave brought to book	那沒有記錄在案的
Who has won a drunken cheer—	備受醉人歡呼的圓滑的政客
The witty man and his joke	專門對準大眾的耳朵揚言
Aimed at the commonest ear,	
The clever man who cries	那些巧言令色的狂喊
The catch cries of the clown,	高唱小丑之歌
The beating down of the wise	棒打智者貶低偉大藝術
And great Art beaten down.	

| 漁夫
—— 葉慈

Maybe a twelve-month since	彷佛一年前
Suddenly I began,	我忽然開始
In scorn of this audience,	藐視這些聽眾
Imagining a man,	設想有這麼一個人
And his sun-freckled face	他滿臉陽光與雀斑
And gray Connemara cloth,	身穿康瑪拉粗衣
Climbing up to a place	從水沫底下的黑石
Where stone is dark with froth,	爬上地面
And the down turn of his wrist	他手腕如常翻動
When the flies drop in the stream—	把魚絲深進水裏
A man who does not exist,	一個不存在世
A man who is but a dream;	如在夢中的人
And cried, "Before I am old	他大聲高呼
I shall have written him one	在我老去之前
Poem maybe as cold	我一定給漁夫寫一首詩
And passionate as the dawn.	如冷靜黎明激發熱情

| **Her Triumph**

—— W. R. Yeats

I did the dragon's will until you came
Because I had fancied love a casual
Improvisation, or a settled game
That followed if I let the kerchief fall:
Those deeds were best that gave the minute wings
And heavenly music if they gave it wit;
And then you stood among the dragon-rings.
I mocked, being crazy, but you mastered it
And broke the chain and set my ankles free,
Saint George or else a pagan Perseus;
And now we stare astonished at the sea,
And a miraculous strange bird shrieks at us.

| 她的勝利
—— 葉慈

你到來之前我一直順從天龍

幻想愛情是逢場作戲

互相妥協或者隨緣指定

發生在放下手帕的遊戲

那事兒叫人一時高飛

或由天範之曲巧妙和鳴

你驟然出現在天龍界裏

我在調侃中為你着迷

但你解開愛的鎖鏈還我自由

如聖喬治或那無神的皮休士

我們如今愕然凝視大海

聽那隻怪鳥對我們淒厲尖叫

| **The Nineteenth Century and After**

—— W. R. Yeats

Though the great song return no more
There's keen delight in what we have:
The rattle of pebbles on the shore
Under the receding wave.

| 十九世紀以後
—— 葉慈

縱使偉大詩歌不再回頭

我們仍然滿足於現時所有

卵石在岸邊應聲碰撞

安居在潮汐底下

About the Author

Born in Hong Kong, China, on May 1, 1934. He obtained a Ph.D. degree from the University of Ottawa in 1961 specializing in Education and Psychology. University teaching is his career, with the University of Toronto as home base, retiring as Professor Emeritus.

He has served as expert advisor to 135 public or academic organizations to promote global understanding, cooperation and peace. They include UNESCO, Asia Foundation, World Bank and governments of developing countries. He was appointed Hong Kong, China, Advisor 1994–97.

From April 1974 onwards he had led 12 groups of professors to visit Chinese mainland to foster institutional relations leading to the opening up process in key universities and the modernization of cities.

His publications in Chinese and English include 87 books, 467 scientific reports, and some 600 poems and translations, promoting humanity ideals and Chinese wisdom.

Honours received include Order of Canada, 31 Citizen and leadership citations, 3 Queen's Celebration awards, Fellow of Russian Academy of Social Sciences, and 7 Honorary Doctoral degrees.